$16.00

ASTROLOGY
30 YEARS RESEARCH

****

By

DORIS CHASE DOANE

AMERICAN FEDERATION OF ASTROLOGERS INC.
P O Box 22040
TEMPE, ARIZONA 85282
1979

First Printing 1956
Current Printing 1985
ISBN Number: 0-86690-070-5

Cover design by *Dynamic Symbology* Katalin B. Williams

Published by:
American Federation of Astrologers, Inc.
P.O. Box 22040, 6535 South Rural Road
Tempe, Arizona 85282

Printed in the United States of America

Books by Doris Chase Doane

ACCURATE WORLD HOROSCOPES
ASTROLOGERS' QUESTION BOX
ASTROLOGY AS A BUSINESS
ASTROLOGY RULERSHIPS
DOANE'S 1981-1985 WORLD WIDE TIME CHANGE UPDATE
HOROSCOPES OF THE U.S. PRESIDENTS
HOW TO PREPARE AND PASS AN ASTROLOGER'S CERTIFICATE EXAM
HOW TO READ COSMODYNES
INDEX TO THE BROTHERHOOD OF LIGHT LESSONS
PROGRESSIONS IN ACTION
TAROT CARD SPREAD READER (With King Keyes)
TIME CHANGES IN CANADA AND MEXICO
TIME CHANGES IN THE WORLD
TIME CHANGES IN THE USA
VOCATIONAL SELECTION AND COUNSELING Vol. I and II
ZODIAC: KEY TO CAREER (With C. Peel)

DEDICATION

To all who assist in the evolution of mankind toward a better world; to the workers who have gone before us; to those now with us; and to those who will follow us through the ages, this work is dedicated.

ACKNOWLEDGMENT

My thanks and appreciation for the encouragement of a host of friends and stellarians, who enthusiastically join me in sharing the unfinished research work.

DORIS CHASE DOANE

Foreword

BEYOND HISTORY and far back into the realms of Legend and Tradition, the more enlightened of the human race have ever striven to find the means of uniting the soul with God. A few of the enlightened ones realized that to accomplish such an exalted goal, it would be necessary to discover a method of uniting the soul — through correspondences — with the manifestation of God as evidenced in the Solar System. They early recognized that an individual could only unite with a power by uniting with the manifestation of this power. Out of such purposeful thinking, Astrology was born, and for uncounted ages Astrology was regarded as a Sacred Science.

In the dawning Aquarian Age, whose Key Phrase is I KNOW, positive knowledge must supplant blind belief. The only method to obtain such knowledge known to The Church of Light is research, testing one by one the ancient tenets of astrology. This book is the result of 30 years research into the Golden Key of Astrology, and it is presented with the hope that the ground covered thus far may provide the earnest student with valuable material heretofore unavailable in organized form.

We trust the time is fast drawing to a close when blind prejudice is substituted for impartial investigation. We also trust that those seeking the Light of Truth are gaining sufficient emotional and intellectual adulthood so that they will not cast aside as worthless statistical research, however rudimentary; but that they will use it as a basis for further study.

We are aware that the evolution of knowledge resulting from further research may alter present conclusions to some degree, and we are conducting further research with all of the means available to us. Astrology demands Eternal Research to provide mankind with a progressive tool insuring a more perfect adaptation to the realities encountered in an ever-changing environment

From time to time, Research Bulletins will be issued by The Church of Light, allowing an opportunity for earnest searchers to keep up to date with the work done by the Astrological Research Department.

In presenting the statistics on disease, no attempt has been made to discuss proper precautionary measures of a dietary and mental nature. Nor has the specific thought-treatment been given to alter a variety of discordant conditions in life. Such information is found in the 21 Courses of Brotherhood of Light Lessons, which, collectively, present the teachings of The Religion of The Stars. Any astrological research conducted by independent individuals or by other organized groups has not been evaluated in this work.

As far as we know, this is the first presentation of a work of this particular type. It is not the last word, nor is it the end of anything. Rather, it is a beginning from which we hope to progress continually toward the advancing horizon of perfection.

This book presents material useful to the student, practical to the professional and valuable to all who seek knowledge for the sake of knowledge.

For such Light as we have been enabled to shed upon the subject of Astrology, we are grateful to the Eternal Spirit, the power behind all manifest creation.

EDWARD DOANE
President,
The Church of Light

•

Introduction

IDEAS THAT GO BACK through ancient Chaldea and ancient Egypt to Atlantis and Mu can be explained in terms of the most modern discoveries of material science. Before man had acquired the knowledge of electromagnetism, of the principles of relativity, of the behavior of electrons, or the principles of radio, or before the exhaustive experiments in psychology which have been made in our universities since the beginning of the present century, these ancient ideas could not be explained.

During the Piscean Age just past, the emphasis has been placed upon unquestioning belief. But during the Aquarian Age, to which the world is now adjusting, authority — no matter how venerable it may be — will have no weight. People will no longer accept a fact on faith or authority. Rightly, to avoid being misled by those who would exploit them, as they were exploited in the past through belief accepted without proof, people will demand facts which they can demonstrate and with which they can prove or disprove things for themselves.

Scientists in our universities as yet have been unwilling to make a statistical study of astrology. But in April, 1924, The Church (formerly Brotherhood) of Light Astrological Research Department—under the direction of Elbert Benjamine, founder — began to solicit from as many persons as it could reach, the birth-date, including hour of birth and date the event happened, of people who had experienced some particular disease or event along with a case history of their lives. In the years that followed, timed birth-charts were statistically analyzed, and the analyses were published in a wide number of books and periodicals. These studies form the sound statistical basis on which Standard Astrology firmly rests, and they have proven invaluable to both student astrologers and professionals alike.

Because most of the published reports are now out of print, I have gone through the wealth of material in the files to gather and present in one volume as much information as possible. There have been other research projects conducted other than those appearing in this book, but I have been unable to find the work sheets, reports and papers concerned. Through the years a great many individuals have contributed to this research activity, but none of them should be held responsible for this presentation, which is strictly my own.

It takes little time and effort to learn to erect a birth-chart and calculate the progressed aspects. If you learn to do this, you are in a position to use the charts of your family and others to prove or disprove rules set forth in The Brotherhood of Light Lessons, which have been further elaborated in these statistical studies.

It would be silly for anyone to believe in astrology because The Church of Light says the planetary positions map states of consciousness which manifest as traits of character, or that the later positions of the planets coincide with characteristic events. And it is equally silly for anyone to believe academic individuals who have made no study of astrology when they say it is a fallacy. It is folly to accept astrology or reject it on authority, because with but little effort anyone can test its truth for himself.

As to how and why events come into life coincide with characteristic progressed aspects, no scientific explanation was possible until Einstein had developed and given to the world his Special Theory of Relativity. As explained by C. C. Zain in his "Laws of Occultism", progressed aspects are consistent with, and logically explained by, Relativity. Now it is possible to understand in terms of accepted science — which the ancients could not do — the various facts of astrology.

But far more important than an explanation, now any person of average intelligence through the application of the facts of astrology can prove it for himself. I believe that anyone who will test the statements in this book will become convinced of astrology. Further, I believe the 21 Brotherhood of Light Courses and this book to be the best possible defense against those who attack astrology.

When you have studied and tested the information given here, you will perhaps understand me when I say that I agreed wholeheartedly with C. C. Zain when he said: "The neophyte striving for adeptship can never hope to attain that exalted state, or even to make much progress toward it, so long as he supinely waits for the misfortunes shown by the stars to overtake him, or languorously looks forward to such benefits as are shown to be showered into his lap. There is not one misfortune that comes into his life that, foreseen, cannot be made less severe in its effect upon him, and not one blessing that cannot in some measure be made more bounteous by the proper use of intelligence and initiative."

DORIS CHASE DOANE

Los Angeles, Calif.
September 1956

Contents

CONTENTS

I
Astrological Combinations

FROM TIME TO TIME in all seriousness some people ask: why bother to go to the lengths demanded in precise chart erection? Others argue that it is not necessary to erect a chart at all, that a solar chart or a sun-sign reading can give a person all he needs to know.

Perhaps these people do not realize the stupendous amount of combinations that can be formed between the various astrological factors. In the Hermetic System of Astrology, 12 signs, 10 planets, 12 horoscope houses and 10 aspects are the factors upon which chart erection and delineation are based.

The least possible number of different combinations resulting from these four groups of astrological factors is:

$$5.3937075 \times 10^{68}$$

This staggering figure should give pause to those brushing aside precision. If that total perplexes the imagination, a couple of comparable ones might help the mind encompass the limitless possibilities. Take the highest United States debt occuring in 1946:

$$2.69422099173 \times 10^{11}$$

Or look at another figure a bit more than the debt, yet still comparable to the number of astrological combinations. Based upon the present estimates, here is the total number of atoms in the universe:

$$3.0 \times 10^{70}$$

Associating these figures of the same order of magnitude should give an inkling as to why the Hermetic System of Astrology insists upon precise chart erection and does not consider a solar chart or a sun-sign reading to be a valid index of the individual.

Here is how the least number of astrological combinations would appear: 539,370,750,000,000,000,000,000,000,000,000,-000,000,000,000,000,000,000,000,000,000,000,000,000. That total is infinitely greater than that of the earth's entire population, which is roughly 2,500,000,000.

II

Astrology vs. Fortune-Telling

THIS CHAPTER, strangely enough, attempts to set forth what Scientific Astrology should NOT, in our present state of knowledge, be expected to do.

Scientific Astrology has greatly suffered through its almost indissoluble marriage with ESP. Knowingly or unknowingly, it is almost impossible to read a birth-chart or to give an opinion on what a progressed aspect is likely to bring into the life without exercising in some degree, Extra-sensory Perception. And no one can state the limit of possible information that can be acquired relative to both past and future through ESP.

The Zenith Foundation, after sifting the vast amount of evidence they had gathered in their thirty weeks of research, during which more than a quarter of a million pieces of mail were received and tabulated, found:

"Authenticated personal experiences indicate that time is not a factor in telepathic communication. Possession of the ability to visualize in detail events which have not happened, a phenomenon science calls precognition, seems but slightly less rare than telepathy itself."

To describe scenes or to give a detailed account of events thus foreseen by *Exrta-sensory Perception,* is *Fortune Telling.* Because many persons calling themselves astrologers possess the ability to do this in a marked degree, and because even the most scientific astrologers usually possess such ability in some degree, there has arisen a widespread belief that astrology foretells Specific Events. Such, however, is not Scientific Astrology, but *Fortune Telling.*

Because of the *Fortune Telling* viewpoint of astrology, almost every year witnesses the discovery of some new astrological system of Predicting Events, which its proud originator proclaims will infallibly indicate the precise nature and precise date of each significant event in any person's life. When the astrologers try this system on numerous charts, and find that it does not enable them to predetermine the Specific nature and exact day of even the more important events, most of them drop it and frantically turn to some still more recently developed System in the hope it will enable them to do, with mathematically ascertained planetary relations, what the accomplished clairvoyant often is able to do with *Extra-sensory Perception*.

We can be quite sure that no such *Fatalistically* precise System of astrology ever will be found. If it ever is, it will demonstrate that no man has power to alter his destiny, that he is not morally responsible for his acts, and that his impression of a succession of events is merely his consciousness moving along a fixed time dimension, with no power whatever to change any of its incidents, all of which are unalterably predetermined. Innumerable experiences and observations demonstrate that the universe is laid out on no such *Fatalistic* Plan.

Yet because of long custom in associating ESP *Fortune Telling* with astrology, it will be difficult to get even many astrologers to adopt a *Non-Fortune Telling* View.

Nevertheless, instead of considering events unalterably fixed along the path of time, let us consider the unconscious mind of a child at birth as composed of the mental elements derived from the experiences it has had in lower forms of life. According to their associations these experiences have been organized, some harmoniously and some discordantly, into thought-cells and thought-structures. Thus the birth-chart maps the thought-cells and thought-structures within the unconscious mind of the new-born babe, and the positions of the planets indicate their kind, intensity, and the general direction of their desires.

The prominence of each planet indicates the volume of energy possessed by the corresponding family of thought-cells. Thus the prominence of Mercury maps the vigor of the Intellectual thought-cells, the prominence of the Moon the vigor of the Domestic thought-cells, the prominence of Jupiter the vigor of the Religious thought-cells, and so on, each planet mapping a group of thought-cells. The expression of the ten thought-cell families, corresponding to the ten planets, comprises the sum total of man's behavior.

The signs of the zodiac in which these planets are found in the birth-chart denote that the Intellectual thought-cells, Domestic thought-cells, Religious thought-cells, etc., mapped by the planets, have a basic tendency to express in the manner characteristic of the sign in which they are located.

The department of life involved in the experiences which built these thought-cells before human birth, and the department of life they consequently will affect during human life, is mapped in the chart of birth by the house which the planet rules.

Whether the desires of the dynamic thought-cells symbolized by the planets, and those of the common thought-cells indicated by the houses, are harmonious or discordant to the individual, and how much so, is indicated by the aspects received by the planet mapping the dynamic thought-structures, ruling the house by which the common thought cells are mapped.

And we have every reason to believe, as the result of long years of careful experiment and observation, that it is the *Psychokinetic Power* exercised by the various types of thought-cells and thought-structures of which the unconscious mind, or soul, is composed, that attracts to the individual the conditions and events of his life, and which influence his thoughts and profoundly affect his physical behavior.

Let us not assume, however, that these thought-cells and thought-structures, even when through a progressed aspect they receive a tremendous new supply of energy to use as *Psychokinetic Power,* are omnipotent. They even then possess only so much energy. And with this amount of energy at their command, they are able to do only a given amount of work in overcoming the resistance of environment that they may realize their desires. As most important events that transpire in the life of man are linked up with factors in his physical environment, we can never afford to neglect the amount of resistance offered by the *particular* physical environment of the individual to these Active thought-cells, in a reasonable attempt to predict the nature and importance of the event coincident with a given progressed aspect.

There is a *Fortune Teller* assumption that two birth-charts with identical positions could belong only to persons of the same ability and same fortune. Observation of charts—over 10,000 of which have been analyzed for (a) Birth-Chart and Progressed Constants in statistical studies by the Church of Light Astrological Research Department, and (b) Natal and Progressed Clues in preliminary research in studies of disease, and for (c) subjective research on innumerable total case histories—however, does not agree with this view. The positions of the signs and planets in a chart give a very good map of the factors of the unconscious mind, both as to volume of energy and their harmonies and discords, relative to each other. But the volume of energy mapped by identical planetary positions in the charts of two different individuals often is markedly different.

While it is true that events and abilities in the lives of identical twins often are startlingly the same; in the lives of other twins it is usual to find similar abilities or trends, but that one takes the lead and goes much further than the other; and often their lives differ also in other respects. It is customary to account for this on the ground that different signs rise, but as a matter of record often the same sign rises in the charts of both. Nor could the few degrees difference on the Ascendant account for this divergency of their lives.

Let us not consider that the birth-chart makes the individual what he is; but that the individual is born when the various heavenly configurations are such that, within limits, they flow in the same direction as the energies within his unconscious mind. This does not necessitate that the positions of the planets must give a portrayal of the *Absolute* Amount of energy of each thought-structure they map, or even that there is no margin of latitude between that which is shown in the sky and that which is within the unconscious mind of the individual. It seems merely that relative to the other factors within the unconscious mind, each thought-structure symbolized by a planet, within broad limits, must be somewhat similar to that planet's prominence and relations to the other planets in the birth-chart which in their turn indicate other important mental factors and their relations within the unconscious mind.

While the experiences of creatures lower in the scale of life, furnishing mental factors so mapped, build up desires of definite type and harmony or discord, these which now guide the *Psychokinetic Power* of the thought-cells are not Conditioned (directed toward) to Specific things in the human life. The work of Mandel Sherman and Irene Case Sherman has demonstrated with the exception of a few reactions—swallowing, closing the eye when the cornea is irritated, sneezing, response to deep pressure, loud noise, restraint, and sudden movement such as dropping — that an infant's reactions are all conditioned (given specific direction) by the pleasure or pain accompanying experiences it has had with physical situations after birth.

Granted that even at the time of birth, as well as subsequently, the thought-cells using *Psychokinetic Power* have a selective action upon environment, nevertheless the Specific things that are likely to be found within the environment vary widely with the historical period, the race, the geological area, the family customs and even the educational and social status of the parents. Yet the growing infant and the thought-cells within its unconscious mind can develop desires toward or away from only those Specific things which are to be found within its physical environment.

The thought-cells, no matter how much *Psychokinetic Power* they use, cannot attract into the life of one child the Specific events which are the customary events in the life of another child. Yet it is these Specific events, as the years pass, which come to direct the desires of the various families of thought-cells toward or away from similar Specific events later in life, when these thought-cells acquire, through progressed aspects, sufficient new energy to attract more events.

This means that, given two birth-charts practically identical—and we have very similar charts from Alaska, Equatorial Africa and New York City in our research files—similar thought-cell activity mapped by identical progressed aspects will attract into the life Specific events that are different. They could not be the same because no amount of *Psychokinetic Power* on the part of the thought-cells can bring snowshoe travel to a Jungleman of the Congo, nor a ride on the Subway to an Indian in Alaska.

Furthermore, because of the early Conditioning of the desires of the thought-cells mapped by the third house, the same aspect which attracts an event where his brethren are concerned to an Alaskan Indian may attract a short journey to a Congo Jungleman, and may bring to a New Yorker the deep study of some scientific book. That is, the third house thought-cells of the Indian may have been Conditioned to express their *Psychokinetic Power* chiefly through events involving his relatives; those of the Jungleman may have had special opportunity early in his life to develop travel desires; while the New Yorker may have been so restricted in early life in reference both to brethren and travel, and been reared where scientific talk was prevalent, and with a library at his hand, that his third house (of the horoscope) thought-cells had learned to gain almost all their satisfaction, not through relatives or travel, but through study. Yet no amount of *Psychokinetic Power* developed from the energy of a progressed aspect could have enabled either the Indian or the Jungleman to have read with either pleasure or understanding the scientific book.

Our extensive research on when and why events happen reveals that the event which is attracted by a progressed aspect, regardless of race and geographical location, bears the characteristics of the planets involved in the aspect. It is accompanied by violence or strife if Mars is involved; by labor, loss or hardship if Saturn is an important member; by good will, abundance or extravagance if Jupiter is in the aspect; by suddenness and accomplished through a human agency if Uranus takes part, etc.

We have demonstrated, in 999 instances out of 1,000, the department of life chiefly affected by the aspect is mapped by a house of the birth-chart ruled by one of the planets taking part in the progressed aspect. To the extent the thought-cells mapped by the aspecting planets in the birth-chart are harmonious or discordant, have been modified in this harmony or discord by events and mental attitudes since birth, and whether the progressed aspect is harmonious or discordant, will the event attracted by the *Psychokinetic Power* of the thought-cells be harmonious or discordant.

The importance of the event attracted is determined (1) by the volume of energy of the thought-cells mapped by the aspecting planets in the birth-chart, (2) the amount of Conditioning they have had through similar events since birth, the (3) amount of energy added to them by the progressed aspect, and (4) the lack of resistance to an important event of the nature of the thought-cells, which with this new progressed aspect energy at their command, the thought-cells try to Demonstrate.

The exact time the event takes place is strongly influenced by cumulative planetary energy. That is, when in addition to the major progressions, there are minor progressions and transit progressions to a planet involved in the major progressed aspect, or to a planet ruling the same house of which a planet involved in the major progressed aspect is co-ruler—thus adding energy to the thought-cells mapped by the same house—this increased activity of a given group of thought-cells tends to determine when the event will transpire.

However, at some period before or after there is a maximum accumulation of planetary energy affecting a given group of thought-cells, the physical environment may offer so little resistance that the thought-cells have no difficulty in Demonstrating the event. Or it may be that just at the time when the accumulated thought-cell energy is greatest, the physical environment may offer so much resistance that no amount of *Psychokinetic Power* can then demonstrate the event. The event which is attracted is brought to pass through work accomplished from the inner plane upon the physical environment by the thought-cells. Both the difficulty of the work at a given time, and the amount of energy that can be directed to it, are factors determining when it can be finished.

With sufficient familiarity regarding the previous happenings and influences in the individual's life which have Conditioned his thought-cells to desire Specific things, and regarding all the various factors in the physical environment at the time, the Scientific Astrologer usually can foretell, not only the Department of life which chiefly will be affected by a progressed aspect, but also the Specific Event, and about how important it will be. But more often than not, this complete information about the character, about the way things customarily develop in the different departments of life, and about the various factors present in the physical environment are known only to the individual himself. If he is his own astrologer, he can know not merely what Specific event to expect, but also what to do to cause the event to be more fortunate than it otherwise would be.

Given something about the environmental conditions, such as the age and previous training, and with certain progressed aspects, the probability of attracting a Specific event (marriage, employment, money, to mention a few) is very high. Not invariably, but in a high percentage of cases, for instance, a powerful progressed aspect to the ruler of the fifth house climaxing about a year after marriage results in the birth of a child.

Often, however, especially when knowledge of the details of the physical environment is lacking, Scientific Astrology must content itself with pointing out that there is a predisposition at a given time toward the occurrance of an important event, relating to one of several departments of life. And within each department of human life there are several alternatives. Thus the fifth house rules love affairs, speculation, entertainment and children. And the same progressed aspect to the ruler of the fifth, due to a planet turning retrograde and then again turning direct, may occur in the same woman's chart at 16 years of age, at 25 years of age, and again at 60 years of age. At 16 it may map a love affair, at 25 the birth of a child, and at 60 a flyer on the stock market. The thought-cells associated with the fifth house in each instance demonstrate an event, but they do not demonstrate the birth of a child in a woman's life at 60 years of age.

Furthermore, it may be possible for one man to accomplish something, due to years of practice and technical training, even when he is under discordant progressed aspects; yet it may be quite impossible for another, without such training, even under the best progressed aspects he can ever have.

Without a thorough knowledge of (1) the Special Conditioning of the thought-cells mapped by the planets through events since birth, (2) the volume and trend of the energy added to the thought-cells by other progressed aspects (Rallying Forces) which at the same time are also within one degree of perfect, and (3) the resistance offered by environment to various events ruled by the houses involved in the various progressed aspects which are present, the Scientific Astrologer usually must content himself to state there is a strong predisposition toward one of the several Specific events, without definitely stating which one will occur.

This, however, enables him to give valuable and specific advice on what the individual should not do, and what he should do. And if the individual himself is the astrologer, with his more complete knowledge of past Conditioning and Present Environment, he can learn also to foreknow, usually, the Specific Event which will be attracted, unless he does something about it, and thus can even more effectively take steps to insure that the events which are attracted into his life are those which he desires.

Research is being carried forward to acquire a more perfect knowledge of those astrological factors which powerfully incline to Specific Events. But relative to many happenings, Conditioning of thought-cells through previous events, and the facility of physical environment to respond to one event ruled by one of the houses involved in the aspect, and its resistance to other events ruled by the houses involved, would appear to be the deciding factors determining which one of several possible events will be the one of chief importance attracted by a given progressed aspect.

Properly considered, astrology cannot be fatalistic, for it deals with mental and emotional elements of the subconscious mind and the times and relative amount of stimulation of these elements. Since the factor of free-will can be used by the intelligent person to direct the energies into constructive channels, no matter how discordant the energy (mapped by active planetary aspects) may appear, scientific astrology cannot be considered *Fortune Telling*.

III
Astrological Research Reports

"IT WAS EINSTEIN'S Special Theory of Relativity, followed to its practical and logical conclusions, which led to the discovery of releasing and utilizing atomic energy. And it is this same Special Theory of Relativity followed to its practical and logical conclusions which indicates how inner-plane energies operate and what can be done to cause them to work more to the individual's advantage.

"As university scientists have conclusively demonstrated that time, distance and gravitation on the inner plane have properties radically different than they have on earth, should we expect inner-plane weather to operate according to the same laws weather operates on earth? Einstein's Special Theory of Relativity carried to its logical conclusions indicates that inner-plane weather affects the individual not merely according to his inner-plane constitution, but through certain time-space relationships that bring structural changes within his astral body.

"Astrological energies constitute the inner-plane weather. How this inner-plane weather affects an individual, however, is not dependent upon any theory; for even as time, distance and gravitation properties of the inner plane have been determined experimentally by university scientists, so have the properties of inner-plane weather, and how it works to affect individuals, groups, cities, nations and world affairs been determined experimentally, and through statistical studies carried out in the process of astrological research.

"One of the outstanding influences of inner-plane weather is that when a person, creature or important event is born, it is born at a time when the inner-plane weather tends to coincide with the inner-plane make-up of that which is then born.

Thus does the inner-plane weather at the time of his birth, as mapped by his birth-chart, indicate the predisposition of an individual to develop abilities of a certain type. The planetary positions and aspects, whatever they may be, which indicate such a predisposition are called its birth-chart constants.

"The predispositions indicated by the inner-plane weather conditions at birth never manifest as events or diseases except during those periods when the appropriate thought-cells receive commensurate additional energy from inner-plane weather mapped by progressed aspects.

"Inner-plane weather consists of astrological vibrations in their infinite variety of combinations. Those mapped by progressed aspects enable planetary vibrations to reach and make active certain groups of thought-cells. These thought-cells have desires such as were imparted to them when they were formed and as indicated by the aspects of the planets mapping them in the birth-chart. Such desires are temporarily altered by the planetary energy reaching them through the inner-plane weather mapped by a progressed aspect. And the additional energy thus reaching the thought-cells not only gives them the power to influence the individual's thoughts and behavior, but it also enables them to attract events of the kind they desire into his life.

"By far the most important inner-plane weather is mapped by major progressed aspects. Church of Light statistical research covering the lives of many thousands of persons indicates that every important event of life takes place during the period while a major progressed aspect is present involving planets characteristic of the nature of the event, and which rule the birth-chart house governing the department of life affected. If more than one department of life is pronouncedly affected by the event, at the time it occurs there are always major progressed aspects involving the ruler of each house governing these various departments of life.

"The periods in his life when the individual is likely to experience a specific event, condition or disease toward which he has a predisposition are indicated by certain major progressed aspects. These progressed aspects mapping inner-plane weather conditions which have been found always to coincide with the given event, condition or disease are called its progressed constants.

"Both the birth-chart position and the progressed position of a planet act as terminals for the reception of the planetary energy. Each terminal actually involved in the progressed aspect receives the energy of the progressed aspect in full volume. But unless the progressed aspect is from a major progressed planet to its birth-chart place—in which case there are only two terminals—each progressed aspect has two other terminals not directly involved in the progressed aspect. Each of these two terminals not directly involved in the progressed aspect receives, through the principle of resonance, one-half as much energy as is received by each terminal directly involved.

"It is important to understand that commonly a major progressed aspect has four terminals because our research has determined that each major progressed constant of an event or a disease is always reenforced by a minor progressed aspect heavier than from the Moon to one of its four terminals at the time the event occurs or the disease develops; and that each reenforced major progressed constant of an event or disease is always released by a transit aspect heavier than from the Moon to one of its four terminals at the time the event occurs or the disease develops. And an independent minor progressed aspect is always released by a transit to one of the birth-chart or major progressed terminals influenced by the minor progressed aspect at the time the event takes place."

As statistically indicated in the Astrological Research Report No. 31, the one effective degree orb of a progressed aspect is always valid. "The more closely the planets approach the perfect progressed aspect the greater the amount of energy the temporary stellar aerial in the astral body is capable of picking up, radio fashion, and transmitting to the thought-cells at its direct and indirect terminals, and the more capable these become of influencing events.

"Due to the reenforcement effect of minor progressed aspects to any of the four terminals of the major progressed aspect, to the trigger effect of transit aspects to any of the four terminals of the major progressed aspect, and to the physical environment through which events must come, the important events attracted by major progressed aspects seldom arrive exactly on the date the progressed aspect is perfect. But other things being equal, they are more apt to arrive close to the date the major progressed aspect is perfect than while the aspect is farther removed. Therefore, that the time and nature of the important events which will be attracted into the life—*unless they are forestalled by precautionary actions*—may be estimated in advance, it is essential that the time be known when each major progressed aspect becomes perfect.

"As major progressions are measured by the ratio of the movements of the planets during one apparent solar day releasing energy which causes the chief structural changes within the astral body that takes place during one astrological year of life, the movements and positions of the planets each four minutes after birth indicate the structural changes that take place within the astral body each corresponding day after birth; the movements and positions of the planets each two hours after birth spread the structural changes so shown over each corresponding month of life after birth, and the movements and the positions of the planets each day after birth indicate the structural changes and events attracted during the corresponding year and time of year of life."

— — From *"Progressing the Horoscope,"* by C. C. Zain

The Church of Light Astrological Research Department began in April, 1924, to solicit data, and since that time has collected, erected and progressed—according to this Hermetic System of Astrology—many, many thousands of birth-charts to the time of some given event. The findings of this activity are presented in the condensed research reports that follow.

STATISTICAL REPORTS

1. Movie Actor

Birth-charts of movie stars analyzed	100	100%
Charts with Mars prominent	98	98%
Charts with Uranus prominent	94	94%
Charts with Neptune prominent	89	89%
Charts with Mercury prominent	79	79%
Charts with Venus prominent	74	74%
Fifth House Activity of 100 Movie Stars		
Charts with fifth house active	93	93%
Charts with planet in fifth house	58	58%
Charts with fifth house more discordant than harmonious	43	43%
Charts with fifth house more harmonious than discordant	53	53%
First House Activity of 100 Movie Stars		
Charts with first house active	91	91%
Charts with planet in first house	54	54%
Charts with first house more discordant than harmonious	58	58%
Charts with first house more harmonious than discordant	40	40%

We tabulated the signs in which the Sun, Moon and Ascendant were located in the charts of 100 movie stars; but no sign stood out with sufficient prominence to be significant, and certainly not sufficiently to be a birth-chart constant of movie actors.

An energetic personality, the outward expression of unusually active thought-cells of the type mapped by Mars appears to be a requisite for motion picture work. Long working hours under great pressure also need strong Mars thought-cells in order to withstand fatigue, criticism from highstrung fellow employees, and to ward off discouragement. Mars is prominent in 98% of the charts.

Ability to fascinate and express more than ordinary charm, to portray a part so faithfully that others feel it, using better and better methods of portrayal to sway the emotions, requires a magnetic personality. Ingenuity and originality is mapped by Uranus, which is prominent in 94% of the charts.

The constant seeking of an ideal is shown by the prominence of Neptune in 89% of the charts. These thought-cells also map ability used to promote a project, to get recognition for one's personality, to reproduce a musical symphony, to portray fictional characters; in fact, Neptune prominence indicates the dramatic ability. The actors having Neptune prominent portray a wide variety of roles, while the ability to enact a charac ter played constantly by the same actor does not require the great dramatic talent mapped by Neptune.

Once a star hits the top, he needs to work to stay there. The aspiring star must work to get there. Thus quick wittedness is important. Mercury, mapping the Intellectual thought-cells, is prominent in 79% of the charts. Venus, prominent in 74% of the charts, coincides with beauty, grace and meticulous dress demanded in many roles. The fifth house, department of entertainment in general and the stage and screen, and the first house, mapping the personality and physical body, are two departments of life active in the chart of a movie star.

2. Marriage

Charts analyzed for marriage	215	100%
Charts with progressed aspect to ruler of seventh	215	100%
Planets forming progressed aspects at Marriage		
Total planets making progressed aspects within one degree of perfect at time of mariage	1,374	100%
Venus	261	19%
Mercury	231	17%
Sun	214	16%
Mars	192	14%
Jupiter	116	8%
Uranus	110	8%
Saturn	80	6%
Neptune	70	5%
Pluto	61	4%
Moon	39	3%
Progressed Aspects of Moon at Marriage		
Charts with progressed Moon tabulated	167	100%
Charts with progressed Moon making no aspect	21	13%

Total progressed aspects of Moon	186	100%
To ruler of seventh	63	34%
To Sun	29	16%
To Venus	28	15%
To Mars	28	15%
To Jupiter	21	10%
To Saturn	18	10%
To Mercury	16	9%
To Pluto	14	8%
To Natal Moon	13	7%
To Neptune	12	7%
To Uranus	7	4%

Two things are made plain by these tables: (1) People marry only when both have a progressed aspect to the ruler of the seventh house to give the thought-cells thus mapped more than their ordinary psychokinetic power. (2) *There is no planet of marriage.* When there is a progressed aspect to the ruler of the seventh, a progressed aspect to Venus favors marriage.

What these tables do not show is that in a woman's chart a progressed aspect to Mars, stimulating amativeness, tends even more toward marriage than a progressed aspect to Venus. The endocrine glands of man usually are active enough without the necessity of a progressed aspect to Mars to stimulate their secretions. But many people marry when there is no progressed aspect to Venus and no progressed aspect to Mars operative.

Marriage is not of necessity associated with affection (Venus). In many marriages the most powerful motive is not love. A stronger motive may be found in the desire for companionship, for a home, for financial security, for children, for social prestige, or for business advantages.

In the life of a woman, the importance of the male sex in general is indicated by the prominence of the Sun in her natal chart, and, how fortunately or unfortunately men tend to affect her life is indicated by the aspects of the Sun. In the life of a man the Moon and its aspects tell the importance, harmony or discord associated with women in general.

In order that marriage—which is thus influenced not only by the desires of the seventh house thought-cells, but also by the desires of the thought-cells mapped by the opposite sex planet, and, in so far as affection is concerned, by the thought-cells mapped by Venus—may not be confused with fifth house activities, it should be pointed out that amative relations, in marriage or out of it, as well as pleasures, entertainment and matters relating to the offspring, are mapped by the fifth house thought-cells.

While there must be a progressed aspect involving the seventh house to attract marriage, that aspect may be weak, provided other progressed aspects at the time are sufficiently powerful. It seems to make little difference in relation to the probability of marriage whether the general progressed aspects and the aspect involving the seventh house are harmonious or discordant if together they provide sufficient psychokinetic power. However, it makes a profound difference in the result of the marriage as will be indicated in a future table relative to divorces. There is very little difference between the progressed aspects that to an unmarried person bring marriage, and those that to a married person bring divorce.

Any seventh house event of importance, whether it be marriage, divorce, a law suit or partnership, requires that the thought-cells mapped by this house gain additional psychokinetic energy.

Not only is this so, but because of the birth-chart harmony of the seventh house thought-cells, the Venus thought-cells, and the thought-cells relating to the opposite sex, the progressed aspects which bring to one person a satisfactory and permanent marriage, if they occur in another person's chart who is already married, may bring divorce. This other person may have the three mentioned sets of thought-cells so discordant that he can get along with no one, and almost any progressed aspect which adds unusual psychokinetic power to the thought-cells mapped by the seventh house and those mapped by the ninth house (court action), if he has previously married, will bring divorce.

Nor does an analysis of the type of progressed aspects present at the time when these 215 marriages occurred—100 of which ended in permanent separation or divorce, but a reasonable number of which were considered highly successful as well as permanent—reveal any particular aspect as outstanding.

Aspects. The semi-sextile took first place, then came the conjunction, to be followed by the sextile and trine. The square and opposition were low, but they occurred about half as often as the sextile and trine.

3. Death of Individual

Charts analyzed for death of individual	108	100%
Charts with progressed aspect to ruler of eighth	108	100%
Charts with progressed aspect to ruler of first	108	100%
Charts with progressed aspect to ruler of eighth discordant only	62	57%
Charts with progressed aspect to ruler of eighth harmonious only	23	21%
Charts with discordant progressed aspects only	40	37%
Charts with harmonious progressed aspects only	00	00%

Heaviest Progressed Affliction at Death

Heaviest affliction a discordant progressed aspect between planets not aspecting each other harmoniously at birth	97	90%
Heaviest affliction a discordant progressed aspect between planets aspecting each other harmoniously at birth	11	10%
Heaviest affliction a harmonious progressed aspect between planets aspecting each other discordantly at birth	00	00%
Planets involved in heaviest progressed affliction	216	100%
Sun	55	25%
Mars	43	20%
Uranus	25	12%
Mercury	23	21%
Saturn	23	21%
Pluto	15	07%
Moon	13	06%
Neptune	12	06%
Jupiter	04	02%
Venus	03	01%

Progressed Moon

Charts with progressed Moon considered	108	100%
Charts with progressed Moon discordant only	46	43%
Charts with progressed Moon making no aspect	26	24%
Charts with progressed Moon harmonious only	19	18%
Charts with Moon making both discordant and harmonious progressed aspects	17	16%

In all 108 charts, at the time of death, there was (1) a progressed aspect to the ruler of the eighth, (2) a progressed aspect to the ruler of the first (including aspects made by the progressed Ascendant and to the birth-chart Ascendant), and (3) at least one discordant progressed aspect.

We may conclude that during the period when an individual does not have progressed aspects in his chart which fulfill these three requisites, he is in almost no danger of dying. Saturn in a discordant progression tends to loss, but people frequently die when there is no progressed aspect to Saturn. There is no planet that can rightly be called the planet of death, although the Sun, planet of vitality, more frequently is a member of a progressed aspect at the time of death.

An individual frequently dies where there are both harmonious and discordant progressed aspects operative in his chart; but in each case there was always a discordant aspect present at the time of death. People die only when some of the thought-cells are stimulated by discordant progressed aspects. But at the time both the progressed aspect to the ruler of the eighth and the progressed aspect to the ruler of the first may be harmonious. Another heavy discordant progressed aspect may act as a Rallying Force rousing discordant desires of these thought-cells sufficiently that, under the environmental facilities present, such as old age, they attract death.

More than merely a progressed aspect to the ruler of the first and a progressed aspect to the ruler of the eighth are necessary to indicate a predisposition toward death during a given period. Conditioning and environment must be considered.

4. Death of Individual and Relatives

Charts analyzed for death of individual and relatives	608	100%
Charts with progressed aspect to ruler of eighth	608	100%

Progressed Moon at Death of Individual and Relatives

Charts with progressed Moon tabulated	282	100%
Charts with progressed Moon discordant only	121	43%
Charts with progressed Moon harmonious only	68	24%
Charts with progressed Moon making no aspect	51	18%
Charts with Moon making both discordant and harmonious progressed aspects	42	15%

Heaviest Progressed Affliction at Death of Individual and Relatives

Planets involved in heaviest progressed affliction	1,216	100%
Sun	241	20%
Mars	219	18%
Saturn	147	12%
Uranus	139	12%
Mercury	137	11%
Pluto	105	09%
Neptune	83	07%
Moon	52	04%
Venus	50	04%
Jupiter	43	04%

In all 608 charts, at the time of death of individual and relatives, there was a progressed aspect to the ruler of the eighth. In this series, as in the previous report on death of the individual, the Sun, planet of vitality, more frequently than any other planet was a member of a progressed aspect at the time of death.

As this and the following tables indicate, there is little danger of death affecting the native's life through the passing of a close relative during periods when there is no progressed aspect to the ruler of the eighth.

If the death of a relative does not affect the individual adversely, the progressed aspect to the ruler of the eighth, the progressed aspect to the ruler of the house indicating the relative, and the Rallying Forces, all may be harmonious. If the death is a distinct loss, the one who suffers by the death of a relative has a discordant progressed aspect at the time.

5. Death of Father

Charts analyzed for death of father	100	100%
Charts with progressed aspect to ruler of eighth	100	100%
Charts with progressed aspect to ruler of fourth	100	100%
Charts with progressed aspect to ruler of fourth harmonious only	19	19%
Charts with progressed aspect also to ruler of tenth	64	64%
Charts with discordant progressed aspect to ruler of tenth	53	53%
Charts with progressed aspects to ruler of tenth harmonious only	11	11%

Progressed Aspects at Death of Parent

Charts analyzed for death of parent	200	200%
Charts with progressed aspect to rulers of both fourth and tenth	131	65%
Charts with heaviest progressed affliction to fourth for death of father and to tenth for death of mother	108	54%
Charts with heaviest progressed affliction to tenth for death of father and to fourth for death of mother	23	11%

The tables relative to the death of the father and mother were compiled at a time when as yet we had no special conviction regarding the tenth house ruling the father or the mother. Cancer, the mother sign, natural ruler of the fourth, seemed to indicate the mother to be ruled by the fourth, and the father by the opposite house in the chart, or the tenth. However, as in each of the 100 charts analyzed for the death of the father, there was a progressed aspect to the ruler of the fourth, as well as one to the ruler of the eighth, and as in each of the 100 charts analyzed for the death of the mother (see next table) there was a progressed aspect to the ruler of the tenth as well as a progressed aspect to the ruler of the eighth, we were forced to conclude that the tenth rules the mother and the fourth the father. These facts have been verified in later research where the mother or father has affected an individual's life in other ways.

That the opposite house would show a progressed aspect to it in 65% of the charts is not surprising when it is considered that the death of the father usually profoundly affects the life of the mother, and the death of the mother usually profoundly affects the life of the father. In other words, if the other parent was still living the progressed aspect involving the surviving parent is about what would be expected; for what affects the surviving parent often has a profound influence upon the individual's life.

Often there are also heavy progressed afflictions to both the fourth and the tenth, as well as to the eighth; and one cannot decide the rulership of father or mother merely because the tenth or fourth is more heavily afflicted by progressed aspects. The surviving parent may influence the individual's life more profoundly at this time than the death of the other parent. If only the tenth is influenced by a progressed aspect, the mother is indicated; and if only the fourth is influenced by a progressed aspect, the father is indicated. But in 65% of the cases considered there were progressed aspects to both the ruler of the tenth and the ruler of the fourth at the death of a parent. Furthermore, in 11% of the charts the affliction to the house ruling the surviving parent was more powerful than the house ruling the parent who died. This indicates that often knowledge of the physical conditions of both parents, and knowledge of the progressed aspects operative in their charts, are required to indicate which one is predisposed in the direction of demise during a period when the predisposition is strongly shown in the individual's chart toward losing a parent. (Father, foster-father, and step-father are indicated by the fourth house.)

6. Death of Mother

Charts analyzed for death of mother	100	100%
Charts with progressed aspect to ruler of eighth	100	100%
Charts with progressed aspect to ruler of tenth	100	100%

Charts with progressed aspect to ruler of tenth harmonious only	16	16%
Charts with progressed aspect also to ruler of fourth	67	67%
Charts with discordant progressed apect to ruler of fourth	50	50%
Charts with progressed aspects to ruler of fourth harmonious only	17	17%

Progressed Moon at Death of Parent

Charts with progressed Moon considered	174	100%
Charts with progressed Moon making discordant aspects only	75	43%
Charts with progressed Moon making harmonious aspects only	49	29%
Charts with progressed Moon aspecting ruler of fourth for father and tenth for mother	46	26%
Charts with progressed Moon making both discordant and harmonious aspects	25	14%
Charts with progressed Moon making no aspect	25	14%

Of the 100 charts analyzed for the death of mother, all show a progressed aspect to the eighth house (death) and a progressed aspect to the tenth house (mother).

However, the tenth and eighth houses also rule other affairs in the native's life. Thus to decide that there is a predisposition during a given period for the mother to die, in addition to the individual's chart, the mother's chart should be consulted, and how she is apt to be affected by progressed aspects under the environmental conditions surrounding her.

It should be noted also, as demonstrated by the progressed aspects in the charts of those who have a foster-mother, or a step-mother, that when such a relationship exists the life is importantly affected only when there is a progressed aspect to the ruler of the tenth. In other words, a foster-mother or a step-mother comes under the influence of the same group of thought-cells as does the blood mother.

7. Death of Husband or Wife

Charts analyzed for death of husband or wife	100	100%
Charts with progressed aspect to ruler of seventh	100	100%

Charts with progressed aspect to ruler of eighth	100	100%
Charts with progressed aspect to ruler of seventh harmonious only	10	10%
Charts with progressed aspect to ruler of eighth harmonious only	16	16%

Whenever the mate affects the life of the individual through some important event, there is always a major progressed aspect to the ruler of the seventh. And whenever the death of anyone affects the life of the individual importantly, there is always a progressed aspect to the ruler of the eighth.

However, the mate does not always die when progressions involve both the seventh and eighth. There can be a number of periods in the life when these progressions are present. As the seventh and eighth rule more than the mate and death, respectively, other events may occur at these times.

To ascertain if, during the period when there is a progressed aspect to the ruler of the seventh and one to the ruler of the eighth, there is a predisposition toward death of the husband or wife, the birth-chart and progressed aspects in the chart of the mate should be consulted. The study should include the conditioning and environment of the mate.

8. Death of Brother or Sister

Charts analyzed for death of brother or sister	100	100%
Charts with progressed aspect to ruler of third	100	100%
Charts with progressed aspects to ruler of eighth	100	100%
Charts with progressed aspect to ruler of third harmonious only	19	19%
Charts with progressed aspect to ruler of eighth harmonious only	06	06%

That death affects the life significantly only when the thought-cells mapped by the eighth house receive additional psychokinetic power through a major progressed aspect is proved by the tables given in connection with the death of the individual. The tables analyzing 608 charts of death of individual and relatives also indicate that there is no planet of death.

When a brother or a sister in any way affect the life there will be a major progressed aspect involving the third house.

In this research on deaths, we have noted that whenever any relative other than parent, brother or sister, marriage partner or child (aunt, uncle, cousin, grandparent, or more distant kin) affects the life importantly because of the relationship, rather than because they become a partner or a foster parent, for instance, there is invariably a progressed aspect to the ruler of the third house. Foster and step-relationships come under the rulership of blood relationships. That is, a step-brother is shown by the third house, etc.

How fortunately or unfortunately the life will be affected by the anticipated event cannot be ascertained from the progressed aspect to the eighth house alone. Even if the event is the death of a relative, the manner in which the individual is affected must be determined by an appraisal of all the progressed aspects operative at the time. The table indicates that a brother or sister may die only when progressed aspects to the ruler of the third and the ruler of the eighth are harmonious. But, in such instances, unless at the time there are heavy discordant Rallying Forces, it is found that the individual has no great sense of loss, but has gained through a legacy or in some way through the death.

9. Death of Offspring

Charts analyzed for death of offspring	100	100%
Charts with progressed aspect to ruler of eighth	100	100%
Charts with progressed aspect to ruler of fifth	100	100%
Charts with progressed aspect to ruler of eighth harmonious only	27	27%
Charts with progressed aspects to ruler of fifth harmonious only	10	10%

When son or daughter dies there is a progressed aspect to the fifth house and a progressed aspect to the eighth house.

People often go through unbelievably powerful discordant progressed aspects without passing on. Thus even when a child's chart is heavily afflicted by progressed aspects, and there is a progressed aspect to its eighth house and a progressed aspect to its first house, it should not be considered more than a danger, for the prevention of which precautionary actions should be taken. And it may be counted on that the child will live if at the same time there is not a progressed aspect to the eighth and a progressed aspect to the fifth, in the charts of both parents.

10. Divorce

Birth-charts analyzed of those permanently separated or divorced	100	100%
Charts with ruler of seventh with both discordant and harmonious aspects	86	86%
Charts with ruler of seventh discordant only	14	14%
Charts with ruler of seventh harmonious only	00	00%
Opposite sex planet with both discordant and harmonious aspects	83	83%
Opposite sex planet discordant only	13	13%
Opposite sex planet harmonious only	04	04%
Charts with Venus with both discordant and harmonious aspects	61	61%
Charts with Venus discordant only	34	34%
Charts with Venus harmonious only	05	05%
Charts with all three elements afflicted	92	92%
Charts with two elements afflicted only	08	08%
Charts with one element afflicted only	00	00%

Progressed Aspects When Divorcee Married

Charts analyzed for marriage	100	100%
Charts with only discordant progressed aspects to ruler of seventh	52	52%
Charts with both discordant and harmonious progressed aspects to the ruler of seventh	31	31%
Charts with only harmonious progressed aspects ruler of seventh	17	17%
Charts with only discordant progressed aspect to ruler of opposite sex	44	44%
Charts with only harmonious progressed aspect to ruler of opposite sex	23	23%
Charts with both discordant and harmonious progressed aspects to ruler of opposite sex	19	19%
Charts with only discordant progressed aspects to Venus	19	19%

Charts with both discordant and harmonious progressed aspects to Venus	33	33%
Charts with only harmonious progressed aspects to Venus	15	15%
Charts with discordant progressed aspects predominant	79	79%
Charts with harmonious progressed aspects predominant	03	03%

Progressed Aspects When Divorced or Separated

Charts of those legally divorced	86	86%
Charts with progressed aspect to ruler of ninth when divorced by the court	86	86%
Charts of those permanently separated	14	14%
Charts analyzed for divorce or permanent separation	100	100%
Charts with only discordant progressed aspects to ruler of seventh	39	39%
Charts with both discordant and harmonious progressed aspects to ruler of seventh	38	38%
Charts with only harmonious progressed aspect to ruler of seventh	23	23%
Charts with only discordant progressed aspect to planet ruling opposite sex	36	36%
Charts with only harmonious progressed aspect to planet ruling opposite sex	24	24%
Charts with both discordant and harmonious progressed aspects to planet ruling opposite sex	20	20%
Charts with only harmonious progressed aspect to Venus	26	26%
Charts with both discordant and harmonious aspect to Venus	26	26%
Charts with only discordant progressed aspect to Venus	19	19%
Charts with both discordant and harmonious progressed aspects	85	85%
Charts with only discordant progressed aspects	11	11%
Charts with only harmonious progressed aspects	04	04%

Progressed Moon When Divorced Or Permanently Separated

Charts with progressed Moon considered	100	100%
Charts with progressed Moon making no aspect	41	41%

Charts with progressed Moon discordant only	35	35%
Charts with progressed Moon harmonious only	15	15%
Charts with Moon making both discordant and harmonious progressed aspects	09	09%

Whether the event relates to competitors, to a law suit, to the attitude of the public, to open enemies, to partnership, or to marriage, *if it is important* it comes into the life only during the time there is a progressed aspect to the ruler of the seventh. The thought-cells thus mapped are furnished much more than their ordinary psychokinetic power. But in so far as the seventh house is concerned, there is no noticeable difference in the aspects that bring marriage, divorce, law suit, or other kinds of seventh house events. Nor were the progressed aspects different, in these charts analyzed, for permanent separation than for legal divorce; except that in all cases of divorce there was a progressed aspect to the ninth stimulating the thought-cells that relate to court action.

The tables show that people get a divorce when there is a harmonious progressed aspect to the ruler of the seventh. And when there is both a harmonious and discordant progressed aspect to the ruler of the seventh they seem to divorce almost as readily as when there is only an unfavorable progression to the ruler of the seventh.

In these tables the three elements considered are: the ruler of the seventh, the ruler of the opposite sex (Sun indicating male, Moon female) and the planet of affection (Venus).

Whether a progressed aspect to the ruler of the seventh tends toward harmony where other people are concerned depends upon the desires of the thought-cells mapped by the seventh house. And these desires are not to be ascertained from a consideration of the particular progressed aspect alone. The desires of the thought-cells as indicated by the aspects in the birth-chart, and as later modified by conditioning, may be even more significant than the quality of the energy added to them by the progressed aspect, even though the progressed aspect tends to make them more harmonious or more discordant, as the case may be.

11. Aviator

Birth-charts of aviators analyzed	100	100%
Charts with Neptune prominent	99	99%
Charts with Mars prominent	96	96%
Charts with Uranus prominent	94	94%

Ninth House Activity of 100 Aviators

Charts with ninth house active	98	98%
Charts with planet in ninth house	57	57%
Charts with ninth house more discordant than harmonious	56	56%
Charts with ninth house more harmonious than discordant	40	40%

Third House Activity of 100 Aviators

Charts with third house active	95	95%
Charts with planet in third house	41	41%
Charts with third house more discordant than harmonious	67	67%
Charts with third house more harmonious than discordant	28	28%

A tabulation in which the signs of the Sun, Moon and Ascendant of these 100 charts were considered revealed that no sign was outstanding enough to warrant special mention; thus no sign can be considered a birth-chart constant of aviators. However, in other research studies we came to the conclusion that Neptune rules aviation.

A prominent Neptune attracts the individual to flying. A prominent Mars maps the required general mechanical ability and the necessary daring in this vocation. And the ability to handle various electrical devices and to use ingenuity in making repairs on delicate instruments is mapped by a prominent Uranus.

One who travels much needs more than ordinary psychokinetic power stimulating the third and ninth houses. If these houses are inactive, he will travel little. As most aviation work involves journeys, to attract to him such trips, the aviator has active third or ninth house thought-cells.

12. Long Journeys

Charts analyzed for long journeys	100	100%
Charts with progressed aspect to ruler of ninth	99	99%
Charts with progressed aspect to ruler of third	53	53%
Charts with progressed Moon making some aspect	79	79%

What would be a long journey to one person might be a short journey to another. The 100 tabulated above were considered long journeys by the native. The one in which a progressed aspect to the ruler of the ninth is missing is one of three trips taken from New York to California and the return by the same individual. At the time of the journey in question there were exceptionally powerful progressed aspects involving the ruler of the third, and progressed Moon was making an aspect to the ruler of the ninth. At the time of the other two trips, there were progressed aspects to the ruler of the ninth.

A progressed aspect to Mercury will not bring travel unless at the same time there is a progressed aspect to the ninth or third house; but when such an aspect to the ruler of travel is present, a progressed aspect to Mercury and to a less degree a progressed aspect to Uranus is an additional testimony of a journey.

A progressed aspect to Neptune will not bring a voyage unless at the same time there is a progressed aspect to the ruler of the ninth house; but when the ninth house ruler is aspected, Neptune—and to a less degree Jupiter—inclines toward the sea.

Before deciding that a progressed aspect to the ruler of the ninth predisposes toward a long journey, or a journey by sea, the birth-chart should be appraised in reference to the activity of the ninth house thought-cells, the past conditioning in reference to long journeys should be determined, and the probable influence of environment during the time the progressed aspect is perfect.

However active the ninth house thought-cells may be, as indicated by powerful aspects to the ruler of the ninth, or many planets in the ninth, the individual may never take a long journey. The ninth house thought-cells early in life may get the habit so strongly of expressing their energy in a specific manner —through publishing, through the practice of law, through religion, or through teaching or other public expression of opinions —that they invariably express through this particular channel whenever they gain new psychokinetic energy from a progressed aspect.

13. Short Journey

Charts analyzed for short journeys	100	100%
Charts with progressed aspect to ruler of third	100	100%
Charts with progressed aspect to ruler of ninth	68	68%
Charts with progressed Moon making some aspect	87	87%

What would be a short journey to some people would be a long journey to others. The journeys tabulated above were considered short by those who experienced them. While progressed aspects to the travel planets, Mercury and Uranus, undoubtedly make travel more certain, the above table indicates that the prediction of a short journey should never be made from aspects to these planets alone; for unless at the same time there is a progressed aspect to the third, the individual will not take a trip of sufficient importance to consider it an outstanding event.

On the other hand, it should be pointed out that people do not take a short journey every time there is a progressed aspect to the ruler of the third house. In early years the individual may have become conditioned so strongly toward expressing through events in which the relatives affect the life, through writing, or through studies, that each time third house thought-cells receive additional psychokinetic energy from a progressed aspect, they desire and work to bring about an event of this particular type. Nevertheless, although the table indicates nothing of the advantages or disadvantages which arose from the journeys, it is quite clear that short journeys of any significance only occur when there is a progressed aspect to the ruler of the third.

14. Gain Money

Charts analyzed for gain of money	100	100%
Charts with progressed aspect to ruler of second	100	100%
Charts with progressed aspect to ruler of second harmonious only	49	49%
Charts with progressed aspect to ruler of second discordant only	24	24%
Charts with progressed aspects to ruler of second both harmonious and discordant	27	27%
Charts with progressed aspect to ruler of second harmonious between planets aspecting each other discordantly at birth	07	07%
Charts with progressed aspect to ruler of second discordant between planets aspecting each other harmoniously at birth	14	14%
Charts with harmonious progressed aspects predominant	76	76%
Charts with discordant progressed aspects predominant	24	24%

Progressed Moon

Charts with progressed Moon considered	54	100%
Charts with progressed Moon harmonious only	19	35%
Charts with progressed Moon discordant only	15	28%
Charts with Moon making both harmonious and discordant progressed aspects	16	30%
Charts with progressed Moon making no aspect	04	07%

Whether or not an individual can gain money during a certain period, and whether at that time it can be acquired from a given venture, are important considerations which can not be determined from the progressed aspects to the ruler of the second house alone, or from the progressed aspects as a whole which are then operative.

The normal ability of the individual to make money is determined by the activity and harmony of six groups of thought-cells: Those mapped by (1) the second house, (2 and 3) the business planets, Jupiter and Saturn, (4) the tenth house, (5) the Sun and (6) the Moon.

In considering the possibility of making money from some

special source or venture, however, the normal ability to gain from that source, indicated by the birth-chart, must be given special consideration: If through business, harmony and activity of the tenth house thought-cells; if preparation and serving of food, the sixth house thought-cells; and so on for the different departments of life indicating the various sources of financial gain.

Making money at a certain time is not merely the result of progressed aspects. It is chiefly determined by the amount of psychokinetic power which special groups of thought-cells acquire, and the desires they have at that time. If they normally express in other ways than gaining money (indicated by heavy birth-chart discords) no amount of harmonious progressed activity can change them enough to demonstrate monetary gain. On the other hand, if these groups of thought-cells normally have a strong desire for money—indicated by harmonious birth-chart aspects, it may be that any progressed energy added to them will be unable to cause them to desire financial loss; then the individual will make some gain even under progressed afflictions.

Nevertheless, the table reveals one thing unmistakably: The other thought-cells, however well intentioned, are powerless to demonstrate money in more than normal amounts, except during such periods as the second house thought-cells gain unusual activity through the energy afforded them by a progressed aspect to the ruler of the second.

15. Lose Money

Charts analyzed for loss of money	100	100%
Charts with progressed aspect to ruler of second	100	100%
Charts with progressed aspect to ruler to second discordant only	64	64%
Charts with progressed aspect to ruler of second harmonious only	12	12%
Charts with progressed aspects to ruler of second both harmonious and discordant	24	24%
Charts with progressed aspect to ruler of second discordant between planets aspecting each other harmoniously at birth	12	12%

Charts with progressed aspect to ruler of second harmonious between planets aspecting each other discordantly at birth	04	04%
Charts with discordant progressed aspects predominant	93	93%
Charts with harmonious progressed aspects predominant	07	07%

Progressed Moon

Charts with progressed Moon considered	74	100%
Charts with progressed Moon discordant only	36	49%
Charts with progressed Moon harmonious only	19	26%
Charts with Moon making both discordant and harmonious progressed aspects	05	05%
Charts with progressed Moon making no aspect	14	19%

These tables in which the progressed aspects are classified at the time financial losses occurred demonstrate clearly that if the birth-chart desires of a group of thought-cells are discordant enough, sufficient harmony commonly is not given them by a harmonious progressed aspect to make them work for things advantageous to the individual. In 12% of these losses, the only progressed aspects to the ruler of the second were harmonious, but in 7% of them, the individual suffered financial loss when the total progressed aspects, from which Rallying Forces are derived, were more harmonious than discordant.

A significant factor which the tables do not reveal, however, is that in each of these instances in which harmonious progressed aspects are dominant, these harmonious progressed aspects are to planets strongly discordant in the chart of birth. This means that if a group of thought-cells is mapped in the birth-chart by heavy discords, their basic desires are so strongly set on demonstrating that which is unfavorable to the individual, that the new and abundant harmonious energy furnished them by a trine through progression enables them to become powerfully active, but is unsuccessful in changing the trend of their desires sufficiently that they work to demonstrate an event completely beneficial to the individual.

16. Accident

Charts analyzed for accident	100	100%
Charts with Mars prominent	79	79%
Charts with Uranus prominent	53	53%
Charts with Saturn prominent	44	44%
Charts with progressed aspect to Mars	100	100%
Charts with progressed affliction to Mars	92	92%
Charts with progressed aspect to Saturn	50	50%
Charts with progressed affliction to Saturn	43	43%
Charts with progressed aspect to Uranus	44	44%
Charts with progressed affliction to Uranus	42	42%
Heaviest progressed affliction from Mars	56	56%
Heaviest progressed affliction from Saturn	20	20%
Heaviest progressed affliction from Uranus	19	19%
Charts with progressed aspect affecting first house	100	100%
Charts with progressed affliction affecting first house	91	91%
Progressed Moon		
Charts with progressed Moon considered	66	100%
Charts with progressed Moon discordant only	27	41%
Charts with progressed Moon harmonious only	19	29%
Charts with Moon making both discordant and harmonious progressed aspects	10	15%
Charts with progressed Moon making no aspect	10	15%

These tables may justify labeling Mars as the accident planet; usually Saturn and Uranus are involved. Those who have a marked predisposition toward accident usually have Mars, and at least one of these other two planets prominent in their birth-charts, and often have all three of them outstanding. One of these three planets—especially Mars—prominent indicates some tendency toward accident. When two of them are prominent the predisposition is more marked. And the more prominent they are and the more discordant, the more certain it is, unless precautions are taken, that the individual will attract accidents.

Quite as constant a factor in accidents as a progressed aspect to Mars is a progressed aspect involving the first house. This does not mean merely a progressed aspect to the ruler of the Ascendant or to a planet in the first house, but must include also any major progressed aspect to the natal Ascendant or a major progressed aspect made by the progressing Ascendant.

Within the compartment of the astral body mapped by the first house are the thought-cells that have been built by experiences relating to the physical body and its condition. And unless they become more than unusually active, through energy coming to them from a progressed aspect, illness or death is not apt to come into the life. But even when a harmonious and not very powerful progressed aspect forms an aerial reaching the first house thought-cells, the planetary energy of any other heavier progressed aspect then active may be able to reach them as a Rallying Force giving these thought-cells the energy and desire to attract events that effect changes in the physical body.

17. Typhoid Fever

Birth-Chart Constant

Birth-Charts analyzed	100	100%
Birth-Charts with rather severe afflictions to Mars	96	96%

Progressed Constants

Charts with major progressed aspects (aspects made by progressed Moon ignored) calculated for the time of typhoid fever	100	100%
Charts with progressed aspect to ruler of Ascendant	100	100%
Charts with progressed aspect to ruler of sixth	100	100%
Charts with progressed aspect to Mars	96	96%
Charts with discordant progressed aspect to Mars	82	82%
Charts with progressed aspect to Mercury, or to a planet in Virgo	100	100%
Charts with progressed aspect to Saturn	83	83%
Charts with no progressed aspect to either Mars or Saturn	04	04%
Charts with no discordant progressed aspect	00	00%

18. Pneumonia

Birth-Chart Constants

Birth-charts analyzed	100	100%
Birth-charts with Sun afflicted by Saturn, Neptune, Pluto or the Moon	98	98%

Progressed Constants

Charts with major progressed aspects (aspects made by progressed Moon ignored) calculated for time of pneumonia	100	100%
Charts with progressed aspect to Mars	100	100%
Charts with progressed affliction to Saturn, Neptune, Pluto, Venus or birth-chart Moon	100	100%
Charts with progressed aspect to Mercury or to planet in Gemini	100	100%

19. Long Life to Men

Charts of men past 70 years of age	80	100%
Sun aspecting Mars	59	74%
Sun in harmonious aspect to Jupiter	41	51%
Sun in an angle	38	48%
Moon aspecting Mars	35	44%
Moon in an angle	27	34%
Moon in harmonious aspect with Jupiter	23	29%

Tabulation of Sun-Sign and Rising-Sign

	Sun-Sign	*Rising-Sign*
Aries	5	5
Taurus	9	7
Gemini	6	6
Cancer	12	15
Leo	9	5
Virgo	4	11
Libra	6	5
Scorpio	3	11
Sagittarius	2	6
Capricorn	13	2
Aquarius	6	4
Pisces	5	3

20. Long Life to Women

Charts of women past 70 years of age	70	100%
Sun aspecting Mars	54	77%
Sun at an angle	31	44%
Moon at an angle	28	40%
Sun in harmonious aspect with Jupiter	27	39%

Moon aspecting Mars	18	26%
Moon in harmonious aspect with Jupiter	17	24%

Tabulation of Sun-Sign and Rising-Sign

	Sun-Sign	*Rising-Sign*
Aries	9	2
Taurus	4	2
Gemini	8	4
Cancer	11	7
Leo	5	8
Virgo	8	7
Libra	3	9
Scorpio	1	11
Sagittarius	3	6
Capricorn	4	4
Aquarius	4	8
Pisces	10	2

The analysis of 150 timed birth-charts of people who lived more than 70 years indicates that for long life the best possible aspect is one from Mars to the Sun, and that the best possible position—other than an aspect—is the Sun in an angle. Any aspect of Mars to the Sun signifies vitality. Although if it is a powerfully discordant aspect it may contribute to shortening the life through infection or accident. Also, harmonious aspects from Jupiter to the Sun increase vitality. For sake of comparison it is better to tabulate the findings relative to men in one table and those relative to women in another.

The charts were analyzed by Sun-sign and by Rising-sign. Cancer, either as a Sun-sign or Rising-sign, maps tenacity of life. But in this group Capricorn indicated long life to more men, but not to women. As a Sun-sign, Cancer, Pisces and then Aries ran next highest in the charts of women, with Virgo and Gemini following closely. Scorpio as a Rising-sign was prominent in the charts of both men and women in this group, but ran quite low as a Sun-sign. Of much more importance, however, than Sun-sign or Rising-sign were heavy afflictions as decreasing longevity and the indicated aspects of Mars and Jupiter to Sun and Moon as increasing it.

21. Childbirth

Charts of women analyzed for birth of a child	100	100%
Charts with progressed aspect to ruler of fifth	100	100%
Charts with progressed aspect to ruler of sixth	100	100%
Charts with progressed aspect to ruler of first	100	100%
Charts of men analyzed for birth of a child	34	100%
Charts with progressed aspect to ruler of fifth	34	100%
Charts with progressed aspect to ruler of sixth	34	100%
Charts with progressed aspect to ruler of seventh	34	100%

When a child is born, there is a progressed aspect—heavier than from the Moon—to the ruler of the fifth and sixth houses in the charts of both husband and wife. In the wife's chart there is also a progressed aspect—heavier than from the Moon—to the ruler of the first house. In the husband's chart there is also a progressed aspect—heavier than from the Moon—to the ruler of the seventh house.

People often have progressed aspects to the rulers of the fifth, sixth and first houses at times when no child is born. Instead, the thought-cell activity shown attracts other events ruled by these houses. Whether the progressed aspects are harmonious or discordant has little significance on whether or not a child will be born, but has much significance as to the effect of the event on the life of the native.

22. Infantile Paralysis

Birth-Chart Constants

Birth-charts analyzed	100	100%
Birth-charts with Mars prominent	99	99%
Birth-charts with Uranus prominent	97	97%
Birth-charts with Saturn prominent	96	96%
Birth-charts with Sun prominent	95	95%

Progressed Constants

Charts with major progressed aspects (aspects made by progressed Moon ignored) calculated for time of infantile paralysis	100	100%
Charts with progressed aspect to Mars	99	99%
Charts with progressed apect to Sun	99	99%
Charts with progressed aspect to Uranus	97	97%
Charts with progressed aspect to Saturn	91	91%

23. Surgical Operations

Charts with major progressed aspects (aspects made by progressed Moon ignored) calculated for time of surgical operation	100	100%
Charts with progressed aspect to ruler of first	100	100%
Charts with progressed aspect to ruler of sixth	100	100%
Charts with progressed aspect to ruler of twelfth	100	100%
Charts with progressed aspect of Mars	100	100%
Charts with progressed aspects of Mars both discordant and harmonious	49	49%
Charts with progressed aspects to Mars discordant only	38	38%
Charts with progressed aspects to Mars harmonious only	13	13%

However ill an individual may be, we may say that if there is no progressed aspect of Mars at the time (heavier than a progressed aspect from the Moon) it is highly unlikely he will have an operation. And it is highly unlikely he will have an operation even when there is a progressed aspect of Mars unless there is also a progressed aspect to the ruler of the first (his body), to the ruler of the sixth (health), and to the ruler of the twelfth (confinement and restriction, but more frequently a hospital).

If the other unfavorable progressions are heavy enough, he may have an operation even if the only progressed aspect to Mars is a semi-sextile, a sextile or a trine. The chances of an operation are greater when the only progressed aspects of Mars are discordant. But in almost half the operations, in addition to one or more harmonious progressed aspects of Mars, there was also a discordant Mars progression.

24. Appendicitis

Birth-Chart Constants

Birth-charts analyzed	100	100%
Birth-charts with Uranus prominent	92	92%
Birth-charts with Mars prominent	90	90%

Progressed Constants

Charts with major progressed aspects (aspects made by progressed Moon ignored) calculated for time of appendicitis	100	100%
Charts with progressed aspect to Mars	100	100%
Charts with progressed aspect to Uranus	97	97%

25. Honors

Charts analyzed for attainment of honor	100	100%
Charts with progressed aspect to ruler of tenth	100	100%
Charts with progressed aspect to Sun	100	100%
Charts with progressed aspect to ruler of tenth harmonious only	32	32%
Charts with progressed aspect to ruler of tenth discordant only	07	07%
Charts with progressed aspects to ruler of tenth both harmonious and discordant	61	61%
Charts with progressed aspect to Sun harmonious only	39	39%
Charts with progressed aspect to Sun discordant only	18	18%
Charts with progressed aspects to Sun both harmonious and discordant	43	43%
Charts with harmonious progressed aspects predominant	76	76%

This table indicates that it is quite difficult to get favorable recognition or to attain a political position when there is only a discordant progressed aspect to the ruler of the tenth. It is also unlikely any honor will be attained during those times there is no progressed aspect to the ruler of the tenth. To get favorable public recognition usually there must be at least one harmonious progressed aspect to the tenth house ruler.

The Sun rules both significance and those in places of authority who often must be relied upon to confer the honor. In cases of gaining a position through popular election, aside from the Sun ruling those in authority, it is the planet of politics. For honor to be attracted there must be not only a progressed aspect to the Sun, but also a progressed aspect involving the tenth.

In considering the activity and the harmony or discord of the thought-cells mapped by the tenth house, the degree occupied by the birth-chart Midheaven and the degree occupied by the progressed M.C. are highly significant. Progressed aspects to these points are potent where credit and honor are concerned.

26. Benefit Through Friends

Charts analyzed for benefit through friends	100	100%
Charts with progressed aspect to ruler of eleventh	100	100%
Charts with progressed aspect to ruler of eleventh harmonious only	41	41%
Charts with progressed aspect to ruler of eleventh discordant only	10	10%
Charts with both harmonious and discordant aspects to ruler of the eleventh	49	49%
Charts with harmonious progressed aspects predominant	66	66%

While this analysis covers only the times when a friend in some manner benefited the individual, charts progressed to the time a friend injured or otherwise influenced the individual indicate that the actions of friends bring events of consequence into the life only at those times when there is a progressed aspect involving the ruler of the eleventh house.

People use their influence in behalf of friends significantly only at times when a progressed aspect to the ruler of the eleventh house gives the thought-cells thus mapped additional psychokinetic energy. These people can be distinguished by house position.

The eleventh house and the twelfth house need to be distinguished from the seventh house in that the seventh house represents those who openly stand before the public in favor of the individual or opposed to him; that is, in a manner they identify themselves as partners with him, or as open opponents. Eleventh house and twelfth house persons may or may not publicly proclaim their desire to help or hinder; but they do exert their influence in various private ways.

In the 100 charts analyzed, some were benefited in each of the various departments of life. In each instance, in addition to the progressed aspect to the ruler of the eleventh, there was also a progressed aspect to the ruler of the house governing the department of life from which the benefit came. Some of the individuals received help from friends when the predominant progressed aspects were discordant. These unfavorable progressions had attracted financial difficulties, illness, loss of job, or trouble with the law, and a friend or friends came forward and helped them out of the difficulty. This is, however, more the exception than the rule. Usually friends are of more help when both the predominant progressed aspects and the progressed aspect to the ruler of the eleventh are harmonious.

27. Tuberculosis

Birth-Chart Constants

Birth-charts analyzed	100	100%
Charts with Saturn prominent	100	100%
Charts with Saturn prominent and afflicted	98	98%
Charts with Neptune prominent	100	100%
Charts with Neptune prominent and afflicted	92	92%
Charts with Jupiter prominent	99	99%
Charts with Jupiter prominent and afflicted	89	89%
Charts with Pluto prominent	97	97%
Charts with Pluto prominent and afflicted	79	79%

Progressed Constants

Charts with progressed aspects (aspects made by progressed Moon ignored) calculated for time of tuberculosis	100	100%
Charts with progressed aspect to Saturn	100	100%
Charts with progressed affliction to Saturn	92	92%
Charts with progressed aspect to Jupiter	98	98%
Charts with progressed affliction to Jupiter	81	81%
Charts with progressed aspect to Neptune	93	93%
Charts with progressed affliction to Neptune	74	74%
Charts with progressed aspect to Pluto	88	88%
Charts with progressed affliction to Pluto	57	57%

28. Cancer

Birth-Chart Constants

Birth-charts analyzed	100	100%
Charts with Saturn prominent	99	99%

Charts with Saturn prominent and afflicted	97	97%
Charts with Jupiter prominent	100	100%
Charts with Jupiter prominent and afflicted	95	95%
Charts with Moon prominent	100	100%
Charts with Moon prominent and afflicted	96	96%
Charts with Neptune prominent	100	100%
Charts with Neptune prominent and afflicted	91	91%

Progressed Constants

Charts with progressed aspects (aspects made by progressed Moon ignored) calculated for the time of cancer	100	100%
Charts with progressed aspect to Saturn	100	100%
Charts with progressed aspect to Jupiter	100	100%

29. Reenforcement Effect of Minor Progressed Aspects

Charts with Major progressions considered	200	100%
Charts showing reenforcement of minor progressed aspects	200	100%
Charts with major progressed constants at time of event	100	100%
Charts with constants reenforced by minor progressions	100	100%
Charts with major progressed constants at time of disease	100	100%
Charts with constants reenforced by minor progressions	100	100%
Charts used in the event series		100
Accidents with major progressed constants involving *Mars* and the *first house*	10	
Marriages with major progressed constants involving *seventh* and *ninth houses*	10	
Divorces with major progressed constants involving *seventh* and *ninth houses*	10	
Long Journeys with major progressed constants involving *ninth house*	10	
Surgical Operations with major progressed constants involving *Mars*, the *first*, *sixth* and *twelfth houses*	10	

Births of Children to women with progressed constants involving the *first, fifth* and *sixth houses* 10
Gains of Employment with progressed constants involving the *tenth* and *sixth houses*.......... 10
Attainments of Honor with progressed constants involving *Sun, Midheaven* and *tenth house*.... 10
Gains of Money with progressed constants involving the *second house*.......... 10
Deaths of Individual with progressed constants involving the *eighth house* and *Ascendant,* or *first house*.......... 10

Charts used in disease series.......... 100
Different diseases picked at random from files.... 100
Charts with specific constants involved.......... 100
Total number of charts used for analysis.......... 200

In every one of these 200 charts studied, each progressed constant was reenforced by a minor progressed aspect. As far as our statistical studies have gone, this reenforcement effect seems to have no influence whatever on the harmony or discord of the major progressed aspect. But it does apparently step up whatever power the major progressed aspect has at that time by the same percentage a similar major progressed aspect steps up the birth chart power of the planets involved.

As one day of major progression time is equivalent to one year of calendar time, major progressed aspects are timed by the apparent movement of the Sun. And as one astrological month of minor progression time is equivalent to one year of calendar time, minor progressed aspects are chiefly timed by the apparent movement of the Moon.

Symbolically, the Sun is the father and the Moon the mother; "male and female created He them." In higher forms of life it requires the union of male and female to bring forth offspring. The Hermetic Axiom, "As it is above, so it is below," is thus verified still further by the finding that both Sun-measured progressed aspects and Moon-measured progressed aspects influencing all its progressed constants must join to give birth to a major event.

From statistical findings we may be justified in assuming that the major progressed constants of an event or disease are always reenforced by a minor progressed aspect (heavier than from the Moon) to one of its four terminals at the time the event occurs or the disease develops. Both birth-chart and progressed positions are terminals and, unless a planet aspects its natal place (involving two terminals) each progressed aspect has four terminals.

NOTE: In several other research studies, at the time of the major event, the major progressed aspect indicating it was always stimulated by a minor progressed aspect and a transit progressed aspect within one degree of perfect.

30. Importance of the Parallel Aspect

Charts progressed to time of attaining honor	100	100%
Charts with progressed aspect involving Sun	100	100%
Charts progressed to the time of surgical operation	100	100%
Charts with progressed aspect involving Mars	100	100%

Because progressed parallels are such long-time aspects, there has been a tendency to underestimate their power, just as there has been a tendency to under-estimate the power of long-time zodiacal progressed aspects. Often zodiacal aspects between slow-moving planets which are within one degree of orb at birth remain in progressed aspect throughout life. Because they (of themselves) bring no sharp change into the life, they may be overlooked. But that the progressed parallel aspects are effective may be observed when other sharper progressed aspects are present. The long-time progressed aspect is reenforced by a minor progressed aspect to one of its four terminals and released by a transit aspect to one of its four terminals. Events governed by a house ruled only by one of the planets involved in the progressed parallel or other long-time progressed aspect often take place.

In early research, we were inclined to believe that events indicated by Mars or the Sun occasionally occurred while the progressed aspect was slightly more than one degree from perfect, thus apparently providing exceptions to the rule. This, however, was misleading.

The first report in which all parallel aspects were considered dealt with attaining honor. It was demonstrated that when the parallels are carefully noted important Sun events take place only when a major progressed aspect involving the Sun is within one degree of perfect. When the report on surgical operations was first issued, one of the charts did not show a major progressed aspect within the one degree orb of perfect involving Mars. However, when this chart was rechecked, it was found that progressed Pluto was well within the one degree orb of parallel birth-chart Mars.

Another instance of the importance of the parallel is a chart on file which is correctly timed at birth, yet showed no zodiacal progressed aspect at the time of death of the individual. When the parallels were considered, the pictured energy was present in sufficient volume to attract the event.

Research into Weather Predicting also demonstrated the importance of the parallel aspect. Without its consideration, errors in gauging and timing weather changes occur. (See next report for additional information.)

31. One Degree Orb Valid for All Progressed Aspects

Charts with progressed constant (Mars) within one degree of the perfect aspect at time of automobile accident	100	100%
Charts with progressed constant (Mars) within one degree of the perfect aspect at time of other accidents	100	100%

In the auto accident charts where the zodiacal progressed aspect involving Mars was more than one degree from perfect, a progressed parallel involving Mars was found well within the one degree of perfect in each case. Other accidents were considered in early research, and at the time only 78% of the charts showed an active Mars involved at the time of the event. However, these calculations were made before there was any ephemeris of Pluto. A recheck of the charts with this later information revealed that the other 22 charts of the series also had a progressed aspect involving Mars which was within one degree of perfect. Pluto's position or parallel aspect accounted for them.

The one degree orb, therefore, may be considered valid. But it was found (occasionally with other planets than Sun and Mars) that in judging, a particular event cannot take place when the relevant progressed aspect is a little beyond the one degree of orb. Therefore, it is essential to ascertain carefully that all parallel aspects by progression involving the planet are also more than one degree from perfect. At times the event indicated by the zodiacal aspect will take place while the zodiacal progressed aspect is as much as a degree and a half from perfect. This, however, must be attributed directly to the effect of the parallel aspect (calculated by declination) within the one degree orb.

32. Influenza

Birth-Chart Constants

Charts analyzed for influenza	100	100%
Charts with Mars prominent	97	97%
Charts with Neptune prominent	86	86%

Progressed Constants

Charts with progressed aspects (aspects by progressed Moon ignored) calculated for date influenza developed	100	100%
Charts with progressed aspect involving Mars	100	100%
Charts with progressed aspect involving Mercury	100	100%
Charts with progressed aspect involving Neptune	98	98%

33. Heart Trouble

Birth-Chart Constants

Charts analyzed for heart trouble	100	100%
Charts with Mars prominent	98	98%
Charts with Sun prominent	96	96%
Charts with Venus prominent	93	93%

Progressed Constants

Charts with progressed aspects (aspects made by Moon ignored) calculated for date heart trouble developed	100	100%
Charts with progressed aspect involving Sun	100	100%
Charts with progressed aspect involving Mars	99	99%
Charts with progressed aspect involving Venus	98	98%
Charts with one or more heavy discordant progressed aspect	100	100%

34. Scarlet Fever

Birth-Chart Constants

Charts analyzed for scarlet fever	100	100%
Charts with Venus prominent	99	99%
Charts with Mars prominent	96	96%
Charts with Neptune prominent	96	96%
Charts with Mercury prominent	95	95%

Progressed Constants

Charts with progressed aspects (aspects made by Moon ignored) calculated for date of scarlet fever	100	100%
Charts with progressed aspect involving Venus	100	100%
Charts with progressed aspects involving Mercury	100	100%
Charts with progressed aspect involving Mars	99	99%
Charts with progressed aspect involving Neptune	97	97%

35. Judging a Contest Winner by Horary Astrology

Charts erected to pick winner of football game	100	100%
Charts for winner of event judged correctly	75	75%
Charts with tenth house predominately harmonious	75	75%
Charts for loser of event judged correctly	25	25%
Charts with tenth house predominantly discordant	25	25%

The winner of a contest might refer to a nation which would win a war, the winner in a prize fight, of an election, of an oratorical contest, of a radio or TV contest, of a horse race, of some type ball game, or the winner of other varied activities.

Research has convinced us that in judging a mental event, (formulating a definite, precise question about one single matter in which the person asking the question is deeply interested and strongly desirous of an answer) the same natal astrological rules apply. While the question is being formulated the mental event is under gestation, even as a child undergoes a period of gestation. When the question is asked, it manifests externally moving from the mental to the physical world even as a child at birth moves into the external world from the womb of its mother. Thus, from an astrological point of view, the birth of a question or idea is the moment when it is expressed verbally or in writing.

A chart erected for the time when such a mental event is

born will reveal as much about it (as we found by studying numerous horary charts with notations of what actually took place relative to the matter asked about) as the chart of a child will reveal about its character and the events it may attract. Within the time limit of the chart—its life—the progressed aspects will indicate what will happen and when it will happen in precisely the same way, and by exactly the same rules.

In a horary question, it is essential that the individual shall have an intense desire to know the answer about some one thing. More than one question asked at the same time brings a multiple birth of mental events, resulting in imperfect births. If there is not an intense desire to know the answer, the mental event does not have sufficient vitality to live, and therefore the expected results are not apt to be realized.

In games in which there was little interest, the result obtained in picking the winner was just about what would be attained through chance. But where there is much interest, the question should be asked relative to the team which one desires to win, or about which at least there is the most interest. Do not ask: Will such and such a team beat such and such a team? Instead, ask: Will such and such a team be the winner?

In handling the 100 football game charts, the question was asked if a certain team would be the winner before the game was played. The chart was erected and judgment passed as to whether or not the mentioned team would win. The harmony and discord of the tenth house was tallied to see which predominated. All the aspects involving planets in tenth, Midheaven and the ruler of the tenth cusp were considered. The astrodynes of all the harmonious aspects thus influencing the tenth house were added together; then the astrodynes of all the discordant aspects thus influencing the tenth house were totaled. If there were more harmodynes than discordynes in the particular chart in question, the team named was judged to win, but if the total discordynes was higher, it was judged that the team would lose.

In winning a game, a prize fight or an election, the main objective is the honor to be gained, although money or other considerations may be of secondary importance. Therefore, the success likely to be attained is judged by the harmony of the house of honor and business (tenth).

36. Nervous Breakdown

Birth-Chart Constants

Charts analyzed for nervous breakdown	100	100%
Charts with Mercury prominent	96	96%
Charts with Uranus prominent	96	96%
Charts with Moon prominent	93	93%

Progressed Constants

Charts with progressed aspects (aspects made by Moon ignored except in next to last statistic) for date of nervous breakdown	100	100%
Charts with progressed aspect involving Mercury	100	100%
Charts with progressed aspect involving Uranus	100	100%
Charts with progressed aspect to birth-chart Moon	89	89%
Charts with progressed aspect to birth-chart Moon or from progressed Moon	100	100%
Charts with one or more heavy discordant progressed aspect	100	100%

37. Automobile Accidents

Birth-Chart Constants

Charts analyzed for automobile accidents	100	100%
Charts with Saturn prominent	96	96%
Charts with Mars prominent	93	93%
Charts with Uranus prominent	91	91%

Progressed Constants

Charts with progressed aspects (aspects made by progressed Moon ignored) calculated for date of automobile accident	100	100%
Charts with progressed aspect involving Mars	100	100%
Charts with progressed aspect involving Mercury	100	100%
Charts with progressed aspect involving ruler of third house	100	100%
Charts with progressed aspect involving Saturn	99	99%
Charts with progressed aspect involving Uranus	99	99%

The birth-chart constants of accidents are Mars, Saturn and Uranus. The more prominent and afflicted they are the greater the predisposition toward accidents.

The progressed constants of accidents, whether to one's person or possessions, or being involved without damage to oneself or one's property, are aspects, especially afflictions, involving Mars, Saturn and Uranus, with special emphasis on Mars. One does not have an accident of importance except when there is a major progressed aspect (heavier than from the Moon) involving Mars. And if at the same time there is a progressed aspect involving Saturn and one involving Uranus the probability of accident is increased.

But when the accident concerns an automobile—whether getting knocked down by one while crossing the street or while riding a bicycle, or being in an auto driven by another, or having one's car smashed by another car, or doing damage to another's person or property while driving one's own car—there are always two other progressed aspects.

Mercury is the planet of travel. The third house is the house of short journeys. In all 100 instances analyzed and studied some automobile was in movement, traveling when it became involved in the accident; and there was present a progressed aspect involving both Mercury and the third house.

38. Tonsil Trouble

Birth-Chart Constants

Charts analyzed for tonsil trouble	100	100%
Charts with Pluto, Neptune or Saturn prominent	100	100%
Charts in which Venus has a discordant aspect	99	99%

Progressed Constants

Charts with progressed aspects (aspects made by Moon ignored) calculated for the date tonsil trouble developed	100	100%
Charts with progressed aspect involving Pluto, Neptune or Saturn	100	100%
Charts with progressed aspect involving Mars	100	100%
Charts with progressed aspect involving Venus	99	99%
Charts with discordant progressed aspect involving Venus	83	83%
Charts with one or more heavy discordant progressed aspect	100	100%

39. Paralysis

Birth-Chart Constants

Charts analyzed for paralysis	100	100%
Charts with Uranus prominent	98	98%
Charts with Mars prominent	97	97%
Charts with Mercury prominent	96	96%

Progressed Constants

Charts with progressed aspects (aspects made by Moon ignored) calculated for date paralysis developed	100	100%
Charts with progressed aspect involving Mercury	100	100%
Charts with progressed aspect involving Mars	98	98%
Charts with progressed aspect involving Uranus	97	97%

40. Eye Trouble

Birth-Chart Constants

Charts analyzed for eye trouble	100	100%
Charts with Mars in discordant aspect to a planet, Midheaven, or Ascendant	100	100%
Charts with Moon in discordant aspect to a planet, Midheaven, or Ascendant	98	98%
Charts with Mars in some aspect with Moon	70	70%
Charts with Mars in discordant aspect with Moon	56	56%

Progressed Constants

Charts with progressed aspects (aspects by Moon ignored) involving Mars at the same time there was some discordant aspect	100	100%

41. Ear Trouble

Birth-Chart Constants

Charts analyzed for ear trouble	100	100%
Charts with Saturn in discordant aspect to a planet, M.C., or Ascendant	100	100%
Charts with Saturn in some aspect with Moon	71	71%
Charts with Saturn in discordant aspect with Moon	50	50%

Progressed Constants

Charts with progressed aspect (aspects made by Moon ignored) involving Mars, at the same time there was some discordant aspect	100	100%
Charts with progressed aspect involving Saturn	99	99%

42. Kidney Trouble

Birth-Chart Constants

Charts analyzed for kidney trouble	100	100%
Charts with Venus afflicted	99	99%
Charts with Mars afflicted	97	97%
Charts with Pluto afflicted	96	96%

Progressed Constants

Charts with progressed aspects (aspects made by Moon ignored) calculated for date kidney trouble developed	100	100%
Charts with progressed aspect involving Venus	99	99%
Charts with progressed aspect involving Mars	99	99%
Charts with progressed aspect involving Pluto	100	100%

43. Diphtheria

Birth-Chart Constants

Charts analyzed for diphtheria	97	100%
Charts with Mars afflicted	97	100%
Charts with Venus afflicted	96	98%
Charts with Pluto afflicted	96	98%
Charts with Mercury afflicted	95	97%
Charts with Neptune afflicted	95	97%

Progressed Constants

Charts with progressed aspects (aspects made by Moon ignored) calculated for date diphtheria developed	97	100%
Charts with progressed aspect involving Mars	97	100%
Charts with progressed aspect involving Mercury	97	100%
Charts with progressed aspect involving Venus	96	98%
Charts with progressed aspect involving Neptune	95	97%
Charts with progressed aspect involving Pluto	94	96%

44. Female Trouble

Birth-Chart Constants

Charts analyzed for female trouble	100	100%
Charts with Mars afflicted	100	100%
Charts with Venus afflicted	97	97%
Charts with Pluto afflicted	97	97%

Progressed Constants

Charts with progressed aspects calculated for date female trouble developed	100	100%
Charts with progressed aspect involving Mars	100	100%
Charts with progressed aspect involving Venus	99	99%
Charts with progressed aspect involving Pluto	99	99%

45. Imprisonment

Charts analyzed for imprisonment	100	100%
Charts with progressed aspect to ruler of twelfth	100	100%
Charts with progressed aspect to ruler of ninth	100	100%
Charts with progressed aspect either to Neptune or Pluto	100	100%
Birth-charts with active twelfth house	69	69%
Charts with progressed aspect to Neptune	95	95%
Charts with progressed aspect to Pluto	90	90%
Charts with discordant progressed aspect predominant	100	100%

Standing out significantly in these charts: (1) the highly active twelfth house in each birth-chart, (2) at the time of imprisonment, the predominantly discordant progressed aspects. While most of these people were imprisoned because of crimes, some were imprisoned due to political or religious ideals contrary to the ruling authority, and some taken captive in time of war.

Whether the imprisonment is due to martyrdom or to crime, the birth chart always shows active twelfth house thought-cells. Except for nurses, doctors, detectives, those working from behind the scenes, or in occult work, strongly active twelfth house thought-cells generally prove detrimental to the fortune. While many who have active twelfth house thought-cells never are in danger of imprisonment, they do attract restrictions and disappointments and private enmities.

People do not become imprisoned while all their progressed aspects are harmonious. Very frequently a person commits a crime, and is either undetected or escapes during the time his progressed aspects are harmonious; only to be discovered or captured and imprisoned when heavy progressed afflictions come along in his chart at the same time there is a progressed aspect to the ruler of the twelfth and ninth houses.

46. Instrumental Musicians

Birth-charts of instrumental musicians analyzed	100	100%
Charts with Moon or Neptune prominent	100	100%
Charts with Mars prominent	88	88%
Charts with Saturn prominent	86	86%
Charts with Moon prominent	86	86%
Charts with Neptune prominent	82	82%
Charts with Venus prominent	79	79%

Fifth House Activity of 100 Instrumental Musicians

Charts with fifth house active	100	100%
Charts with planet in fifth house	41	41%
Charts with fifth house more discordant than harmonious	51	51%
Charts with fifth house more harmonious than discordant	49	49%

In a tabulation of 100 charts of instrumental musicians by the signs occupied by Sun, Moon, and Ascendant, we found that Virgo, Aquarius, and Pisces were highest, with Leo next, Taurus was lowest with more than half as many as Virgo, and Capricorn was next lowest. No zodiacal sign significantly portrays the ability of an instrumental musician.

Prominent Mars thought-cells map creative energy and muscular control; Saturn maps plodding perserverance required for long, monotonous practice sessions; the Moon maps time, tune and rhythm; Neptune maps feeling and dramatic approach; and Venus, while the planet of art, maps a good ear for harmony and tonal quality.

47. Vocal Musician

Birth-charts of vocal musicians analyzed	25	100%
Charts with Moon or Neptune prominent	25	100%
Charts with Mars prominent	24	96%
Charts with Moon prominent	21	84%
Charts with Neptune prominent	19	76%
Charts with Saturn prominent	17	68%
Charts with Venus prominent	16	64%

In tabulating 25 charts of vocal musicians by the sign-position of the Sun, Moon and Ascendant, Leo was highest, with Sagittarius a poor second. Much the lowest was Capricorn, with Aries and Pisces next lowest. Too much reliance cannot be placed on this factor, however, as only 25 accurately timed birth-charts were available.

48. Composer of Music

Birth-charts of composers analyzed	25	100%
Charts with Uranus and Neptune prominent	25	100%
Charts with Uranus prominent	24	96%
Charts with Neptune prominent	23	92%

For originality, Uranus should be prominent in the chart of a composer. Fine feeling and fertility of imagination is mapped by a prominent Neptune. All these great composers' charts showed Uranus and Neptune prominent.

49. Musical Conductor

Birth-charts of musical conductors analyzed	10	100%
Charts with Uranus prominent	09	90%
Charts with Mars unusually prominent	09	90%

Conductors use more Mars creative energy and more ability to influence others through the personal magnetism indicated by a prominent Uranus.

If neither Moon nor Neptune are prominent, chances are poor a person would be successful in the music profession (See Report 46).

50. Horary Astrology

Charts erected for time a question was asked	100	100%
Outcome of question timed by a progressed aspect	100	100%

Progressed aspects proved the best means of timing events in the study of 100 Horary charts where the date was known when the indicated event happened. A major event was timed by a major progressed aspect, a minor event by a minor progressed aspect, and an insignificant event by a transit. At the time of the major event, the major progressed aspect indicating it was stimulated by a minor progressed aspect and a transit just as in Natal Astrology (Hermetic System).

In this system, the Moon is no longer considered as co-ruler of the Querent unless the Moon is in the first house of the chart, or unless Cancer rises or intercepts the first house.

Translation of Light and Collection of Light do not apply when this method is used. No reason was found to avoid judgment when the Moon or a significator was Void of Course. No reason was found to avoid judgment of a horary chart when more than 27 or less than 3 degrees of a sign were found on the Ascendant, if this method of delineation is followed.

51. Doctor

Birth-charts of doctors analyzed	100	100%
Charts with Mars prominent	99	99%
Charts with Jupiter prominent	99	99%
Sixth House Activity of 100 Doctors		
Charts with sixth house active	99	99%
Charts with planet in sixth house	45	45%
Charts with sixth house more discordant than harmonious	51	51%
Charts with sixth house more harmonious than discordant	47	47%
Twelfth House Activity of 100 Doctors		
Charts with twelfth house active	99	99%
Charts with planet in twelfth house	99	99%
Charts with twelfth house more discordant than harmonious	58	58%
Charts with twelfth house more harmonious than discordant	39	39%

In tabulating the signs of the Sun, Moon and Ascendant in the 100 charts of doctors, no sign stood out with sufficient prominence to warrant special mention.

Building and tearing down are activities typical of the Mars thought-cells. Mars is prominent in a doctor's birth-chart maping thought-cells sufficiently active to make a success of an occupation calling for either constructive or destructive ability. These thought-cells also relate to surgical skill, administration of medicine and other methods of repairing the body.

Through the power of suggestion, the mental and personal atmosphere of the doctor is an important factor in healing.

The prominent Jupiter in a doctor's chart maps the warmth, geniality, confidence and optimism communicated to the patient.

As doctors are continuously brought into contact with sickness, the thought-cells mapped by the sixth house are unusually active. While only some of the doctors associate with hospitals, the sick-rooms they visit constantly are places of confinement and restriction to the patient; thus the twelfth house is active in a doctor's chart.

52. Gain Employment

Charts analyzed for getting employment	100	100%
Charts with progressed aspect to ruler of tenth	100	100%
Charts with progressed aspect to ruler of sixth	100	100%
Charts with progressed aspect to ruler of tenth harmonious only	60	60%
Charts with progressed aspect to ruler of sixth harmonious only	46	46%
Charts with harmonious progressed aspects predominant	61	61%
Charts with discordant progressed aspects predominant	39	39%
Charts with progressed aspect to ruler of tenth discordant only	32	32%
Charts with progressed aspect to ruler of sixth discordant only	37	37%
Charts with progressed aspects to ruler of tenth both harmonious and discordant	08	08%
Charts with progressed aspects to ruler of sixth both harmonious and discordant	17	17%

The tenth house thought-cells have been formed in human life by experiences relating their desires and activities to honor, credit, profession and superiors. This means that the job itself, apart from the actual labor involved, is ruled by the tenth house and subject to the tenth house thought-cells.

On the other hand the sixth house thought-cells have been formed by experiences relating to conditions surrounding the work—whatever it is—as well as to labor, foods, small animals, employees, services and illness. Employment involves the job itself, which is ruled by the tenth house, and the services rendered, ruled by the sixth house. This explains why, as revealed by the tables, both tenth house and sixth house thought-cells must gain more than normal activity, through their rulers being involved in progressed aspects, when employment is secured.

53. Lose Employment

Charts analyzed for losing employment	100	100%
Charts with progressed aspect to ruler of sixth	100	100%
Charts with progressed aspect to ruler of tenth	97	97%
Charts with progressed aspect to ruler of sixth discordant only	70	70%
Charts with progressed aspect to ruler of tenth discordant only	65	65%
Charts with discordant progressed aspects predominant	91	91%
Charts with harmonious progressed aspects predominant	09	09%
Charts with progressed aspect to ruler of tenth harmonious only	19	19%
Charts with progressed aspect to ruler of sixth harmonious only	17	17%
Charts with progressed aspects to ruler of tenth both harmonious and discordant	16	16%
Charts with progressed aspects to ruler of sixth both harmonious and discordant	13	13%

In the loss of employment, the environmental factor becomes important. So far as holding a job is concerned this table indicates the sixth house to be more important than the tenth. People do not lose jobs merely because they are discharged, or because they cannot get along with superiors (the tenth house and the Sun); but at times they voluntarily quit for different reasons: the work becomes too difficult; they move to a different locality, etc. But, in getting a job, which depends upon the belief of the employer (tenth) in the ability (reputation) of the employee to fill the job, the tenth house seems to be more important than the sixth.

The 100 charts include those who obtained or lost jobs during the boom years and those who gained or lost jobs in the depression years. It took more sixth and tenth house thought-cell activity in periods of depression to get a job, and a lot less sixth and tenth house thought-cell activity to lose a job in periods of ready employment.

If a person is idle and goes to work, if he is working and loses his job, or if he changes from one job to another, this is an event not only affecting his position, but also affecting the labor and the conditions surrounding it; and thus it involves both tenth and sixth house thought-cell activity. Such activity is present only when these thought-cells acquire additional energy from a progressed aspect to the rulers of the tenth and sixth.

54. Writer

Birth-charts of writers analyzed	100	100%
Charts with Mercury prominent	95	95%
Charts with Mercury aspecting Moon or Sun	91	91%
Charts with Mercury aspecting Moon	64	64%
Charts with Mercury conjunction Sun	58	58%
Third House Activity of 100 Writers		
Charts with third house active	97	97%
Charts with planet in third house	49	49%
Charts with third house more discordant than harmonious	58	58%
Charts with third house more harmonious than discordant	40	40%
Ninth House Activity of 100 Writers		
Charts with ninth house active	94	94%
Charts with planet in ninth house	46	46%
Charts with ninth house more discordant than harmonious	51	51%
Charts with ninth house more harmonious than discordant	49	49%

In tabulating the signs in which the Sun, Moon and Ascendant were located in the charts of the 100 writers, no sign stood out with sufficient prominence to warrant special mention. No sign can be considered a birth-chart constant of writers. The prominent Mercury indicates thought-cells facile in expressing through written and spoken language.

In the five charts where Mercury was not prominent an unusually prominent Uranus or an unusually prominent Moon (writing largely through impression) acted as a substitute for a prominent Mercury. The five charts were those of writers of outstanding recognition.

Writing is an activity associated with the third house, and only those with active third house thought-cells are apt to feel a strong enough attraction to make writing as important as it must be in this vocation. An active ninth house is necessary to get the material published.

55. Teacher

Birth-charts of teachers analyzed	100	100%
Charts with Mercury prominent	84	84%
Ninth House Activity of 100 Teachers		
Charts with ninth house active	96	96%
Charts with planet in ninth house	51	51%
Charts with ninth house more discordant than harmonious	61	61%
Charts with ninth house more harmonious than discordant	38	38%
Fifth House Activity of 100 Teachers		
Charts with fifth house active	92	92%
Charts with planet in fifth house	45	45%
Charts with fifth house more discordant than harmonious	60	60%
Charts with fifth house more harmonious than discordant	39	39%

An analysis of the signs in which the Sun, Moon and Ascendant are found in these 100 charts does not indicate any sign as a birth-chart constant of teaching. However, Gemini, Virgo or Sagittarius seems to attract people to this vocation more than any other sign. Also, every additional aspect to Mercury seems to strengthen the attraction.

A prominent Mercury is commonly found in the chart of a teacher. To the extent Mercury is prominent and strongly aspected is there ability to impart information to others. Numerous Mercury aspects map the facility to grasp and teach a variety of subjects.

Teachers exercise the thought-cells mapped by the ninth house. Any planet in the ninth house and the ruler of the cusp of the ninth house to a less pronounced degree modifies somewhat the manner in which the opinions are publicly expressed.

56. Lawyer

Birth-charts of lawyers analyzed	100	100%
Charts with Mercury or Uranus prominent	100	100%
Charts with Saturn prominent	99	99%
Charts with Mercury prominent	96	96%
Charts with Uranus prominent	95	95%
Charts with Jupiter prominent	89	89%
Ninth House Activity of 100 Lawyers		
Charts with ninth house active	100	100%
Charts with planet in ninth house	53	53%
Charts with ninth house more discordant than harmonious	59	59%
Charts with ninth house more harmonious than discordant	40	40%

An analysis of the signs occupied by the Sun, Moon and Ascendant in the 100 charts gives the lead to Leo, Libra, Sagittarius and Capricorn; with Gemini and Pisces taking low places. The difference between the high and the low, however, is not significant enough to point to any zodiacal sign as a birth-chart constant of lawyers. Nor is it important enough that people with the Sun, Moon or Ascendant in Pisces or Gemini should be discouraged from taking up law, provided the indicated planets are prominent and the ninth house is sufficiently active.

An active mind, alertness, shrewdness, ability to study and ease of expression are mapped by a prominent Mercury or Uranus. The ability to work hard at monotonous work, systematize evidence and reason logically are mapped by Saturn. For patronage in any of the professions, the good-fellowship of the Jupiter thought-cells is required. In the eleven charts where Jupiter was not prominent, the Sun was.

Association with the courts is attracted by thought-cell activity mapped by the ninth house. If these thought-cells are not unusually active, there will be little attraction to legal matters.

57. Politician

Birth-charts of politicians analyzed	100	100%
Charts with Sun prominent	100	100%
Charts with Pluto prominent	97	97%
Charts with Uranus prominent	97	97%
Charts with Mercury prominent	86	86%
Charts with Jupiter prominent	84	84%

Tenth House Activity of 100 Politicians

Charts with tenth house active	100	100%
Charts with planet in tenth house	64	64%
Charts with tenth house more discordant than harmonious	54	54%
Charts with tenth house more harmonious than discordant	44	44%

The seventh house activity in the 100 charts of politicians was tabulated and no activity above or below normal was shown. The signs in which the Sun, Moon and Ascendant were located revealed no emphasis on any sign which can be considered as a birth-chart constant for politicians.

The thought-cells ruled by a prominent Sun are the chief source of a person's ability to win the following of others and to exercise authority. Pluto is the general ruler of dictators, and the dictator type of politician needs active Pluto thought-cells. But even if not given to dictating, anyone who handles groups of people needs these thought-cells active. Ability to handle groups, or large masses, is highly advantageous to anyone aspiring to political honors.

Personal magnetism used by politicians to influence others and influence their minds is mapped by a prominent Uranus. The shrewdness needed by a politician is mapped by a prominent Mercury. The joviality, good-will factor, and fellowship resulting either in patronage or votes is mapped by a prominent Jupiter.

Politics is essentially a tenth house association in two different ways. The tenth house maps the thought-cells which relate to political office and to the administration of public business. But in addition, political choice—even when the exercise of political power is not known to the public at large—depends mainly upon the reputation, which relates to thought-cells mapped by the tenth house. Thus it appears essential for one who takes up politics as a vocation to have an active tenth house.

58. Policeman

Birth-charts of policemen analyzed	100	100%
Charts with Saturn prominent	100	100%
Charts with Mars prominent	100	100%
Charts with Pluto prominent	97	97%
Charts with Neptune prominent	95	95%

Twelfth House Activity of 100 Policemen

Charts with twelfth house active	100	100%
Charts with planet in twelfth house	51	51%
Charts with twelfth house more discordant than harmonious	57	57%
Charts with twelfth house more harmonious than discordant	40	40%

Tabulating the 100 charts according to the signs occupied by the Sun, Moon and Ascendant showed Cancer, Leo, Virgo and Libra more numerous; with about an equal number in each. The fewest had these factors in Aquarius, next fewest in Pisces, with Gemini only a little more numerous than Pisces. Yet policemen are sufficiently numerous with Sun, Moon and Ascendant in Aquarius that no one should be discouraged from taking up this vocation because these factors are in that sign. Nor can any one sign be considered the birth-chart constant of the vocation to the extent that one should be encouraged to take it up on the strength of sign positions.

Routine nature of the work, persistent purpose in carrying out the letter of the law and strict justice are mapped by a prominent Saturn. Mars was prominent in all cases and usually afflicted—more often than not by the Moon—mapping the thought-cells indicating courage and daring to face physical danger. A prominent Pluto maps the association with antisocial individuals and the need to dictate. In order to plan and trap his quarry, a prominent Neptune, as well as an active twelfth house—especially if he is doing secret service work—is an advantage to a policeman.

59. Salesman

Birth-charts of salesmen analyzed	100	100%
Charts with Jupiter or Uranus prominent	100	100%
Charts with Jupiter prominent	92	92%
Charts with Uranus prominent	89	89%
Charts with Mercury prominent	85	85%

An analysis of the sign locations of the Sun, Moon and Ascendant in the charts of 100 salesmen found Leo and Virgo highest, Aries fewest, with Pisces and Taurus making a poor showing. Yet Aries produced more than one-third as many, and Taurus more than one-half as many as the two high signs. Analyzing by Sun-sign only, Leo ranked first, with Gemini next. Aries by Sun-sign alone produced the fewest, with about one-third as many. Yet while Leo seems to be favorable to salesmen, it cannot be considered a birth-chart constant of that vocation. Enough successful salesmen are born under strong influences from all the other signs, and without the more favored signs prominent, that it seems inadvisable to discourage any person from taking up salesmanship merely on the ground that he has Sun, Moon or Ascendant or planets in certain zodiacal signs.

The table, as well as common observation, indicates two distinct types of salesmen: One who depends upon the good-will of his customer to make his sale, and the other depends upon his persuasive powers. Most successful salesmen are to an extent a combination of both types.

Jupiter prominent coincides with various acts of friendship, joviality, and perhaps the ability to tell entertaining stories; all of which gives the impression of the salesman as a fine fellow. He gets good-will. The ability to exercise magnetic control over others and to make decisions quickly is mapped by a prominent Uranus. Mercury prominent maps glibness of tongue and a sharp intellect.

The analysis reveals that unless either Jupiter or Uranus is prominent in the birth-chart, it is better to discourage such an individual from engaging in salesmanship as his vocation. Those most successful in such work have both Jupiter and Uranus conspicuously prominent in their charts of birth.

60. Astrologer

Birth-charts of astrologers analyzed	100	100%
Charts with Uranus prominent	100	100%
Charts with Neptune or Pluto prominent	100	100%
Charts with Mercury prominent	94	94%
Charts with Pluto prominent	92	92%
Charts with Neptune prominent	86	86%

Tabulating the signs occupied by the Sun, Moon and Ascendant, none stood out from the others sufficiently to warrant special attention, and certainly no sign can be considered a birth-chart constant of astrologers.

A person with a prominent Uranus, Neptune and Pluto may be attracted to astrology or occult work. However, these planets prominent are also associated with other types of work. Uranus, planet of astrology, is prominent in the charts of all astrologers. In these 100 charts of astrologers, it was found that those who follow professional astrology with some degree of success almost invariably have a prominent birth-chart Jupiter, mapping the ability to attract patronage. A prominent Mercury maps thought-cells of a studious nature necessary to an astrologer. In the cases where Mercury was not prominent (6%), its higher octave—Uranus—was prominent.

There seem to be three types of astrologers: 1) Those with active Mars thought-cells (usually Mars in close aspect with Mercury) indicating mathematical ability; 2) Those with a dominant Neptune, but with weak Uranus, Mars and Mercury, lacking ability and initiative to work at scientific delineation, depending upon their psychic faculties alone; 3) and The typically scientific Uranian astrologers who do the most valuable work, by combining mathematical and psychical abilities. A prominent Pluto maps the thought-cells which assist the astrologer to understand how the planetary energies operate and what to expect from them.

61. Draftsman

Birth-charts of draftsmen analyzed	100	100%
Charts with Mercury prominent	100	100%
Charts with Neptune or Venus prominent	100	100%
Charts with Mars prominent	97	97%
Charts with Venus prominent	91	91%
Charts with Neptune prominent	88	88%

Third House Activity of 100 Draftsmen

Charts with third house active	100	100%
Charts with planet in third house	58	58%
Charts with third house more discordant than harmonious	70	70%
Charts with third house more harmonious than discordant	30	30%

A tabulation of the signs in which the Sun, Moon and Ascendant are found reveal more draftsmen in Leo and Cancer, with almost as many in Libra and Pisces, while Aries, Taurus and Sagittarius contain fewest, with only two-thirds as many as in the four favored signs. Yet as many successful draftsmen are born under Aries, Taurus and Sagittarius, this does not warrant discouraging a person from taking up the vocation simply because he was born under one of these three signs.

Mercury rules the nerve currents; and if there is to be fine co-ordination between the images in the mind and muscular action, there must be sufficient brain-cell activity and nervous control, which is present only among those who have Mercury prominent in their birth-charts.

Most draftsmanship is employed in the precise plotting of something to gain more ready knowledge of it, or in plans or designs from which something is to be built. This demands a knowledge of mechanics and structural work easily acquired only by those who have Mars thought-cells prominent. Successful drafting requires neat and attractive sheets. Proper lettering is often an important feature. The artistic appearance essential to success in this work is made possible chiefly through active Venus thought-cells, or those of its higher octave, Neptune.

Drafting, like writing, is a third house activity; and if one is to do sufficient work to make a success of this occupation, he must have sufficiently active thought-cells mapped by the third house.

62. Engineer

Birth-charts of engineers (electrical, civil, mechanical, construction, aviation, radio, etc.) analyzed	100	100%
Charts with Uranus prominent	100	100%
Charts with Mars prominent	100	100%
Charts with Saturn prominent	100	100%

Tabulating the signs in which Sun, Moon and Ascendant are found, Leo in the 100 charts of engineers runs about 30% above average; Libra, Scorpio and Cancer about 10% above average; and Pisces about 30% below average. Yet no one should be discouraged from taking up engineering; for engineers with such positions are not uncommon, as this analysis shows.

Some books say that Mars-Uranus aspects map engineering ability. Yet this certainly cannot be considered as a birth-chart constant of engineering, as in 47 of these 100 charts no such aspect was present.

Almost any type of engineering requires ingenuity and ability to solve problems, mapped by a prominent Uranus. Engineering more commonly is employed in association with constructing something. The natural aptitude for any type of creative endeavor or building is largely measured by the amount of Mars activity in the birth-chart.

To become a successful engineer, an individual must be prepared to shoulder responsibility, and to work hard and persistently. He needs system, foresight, and logical reasoning to make successful plans. A prominent Saturn coincides with these qualifications.

63. Farmer

Birth-charts of farmers analyzed	100	100%
Charts with Saturn prominent	100	100%
Charts with Pluto or the Moon prominent	100	100%
Charts with Pluto prominent	92	92%
Charts with Moon prominent	88	88%

Fourth House Activity of 100 Farmers

Charts with fourth house active	100	100%
Charts with ruler of fourth member of heaviest configuration in chart	87	87%
Charts with planet in fourth house	49	49%
Charts with fourth house more discordant than harmonious	58	58%
Charts with fourth house more harmonious than discordant	40	40%

The tabulation of the signs in which the Sun, Moon and Ascendant were found in the 100 charts brought out some quite unexpected data. For instance, it has generally been held that farmers were more apt to have these influences in earthy signs; yet in 45 of these charts the Sun, Moon and Ascendant were in other than earthy signs.

Obviously enough, and also contrary to prevalent opinion, not only Virgo but Libra ran highest, with 40% above average. Perhaps the fact that Libra is the exaltation of Saturn has something to do with its high rating; yet the signs ruled and co-ruled by Saturn — Capricorn and Aquarius — ran low. Virgo, the harvest sign, running high was in conformity with prevalent opinion.

Aquarius ran lowest, with 24% below average, and Taurus, Scorpio and Capricorn next lowest, with 12% below average. Yet Taurus has commonly been considered a farming sign, and Libra a sign of avoiding agriculture.

A prominent Saturn maps the hard and monotonous labor, long hours of work, patience, planning ability and foresight required of the farmer. A prominent Moon maps the farmer's function of feeding the world via crops, raising chickens and breeding livestock. A prominent Pluto, upper octave of the

Moon, coincides with mass production of fruits, grains, vegetables and stock.

Active fourth house thought-cells show the association with land.

64. Nurse

Birth-charts of nurses analyzed	100	100%
Charts with Moon prominent	100	100%
Charts with Mars prominent	100	100%
Charts with Pluto prominent	97	97%
Sixth House Activity of 100 Nurses		
Chart with sixth house active	100	100%
Charts with planet in sixth house	53	53%
Charts with sixth house more discordant than harmonious	62	62%
Charts with sixth house more harmonious than discordant	35	35%
Twelfth House Activity of 100 Nurses		
Charts with twelfth house active	95	95%
Charts with planet in twelfth house	52	52%
Charts with twelfth house more discordant than harmonious	59	59%
Charts with twelfth house more harmonious than discordant	40	40%

A tabulation of the signs in which the Sun, Moon and Ascendant were found showed no variation from normal sign distribution sufficiently important to warrant mention.

The ability to take care of the weak and helpless and look after their wants is mapped by a prominent Moon. Ability to assist the patient rebuild his body, and courage to unflinchingly witness pain and suffering are mapped by a prominent Mars. Pluto, the higher octave of the Moon, maps the desire to serve with pride in relieving the pain of others.

Active sixth house thought-cells attract environments of illness, among other things, and active twelfth house thought-cells are associated with hospitals and places of confinement. As these houses map the type of environment in which a nurse works, the thought-cells of both departments of the chart should be active.

65. Machinist

Birth-charts of machinists analyzed	100	100%
Charts with Mars prominent	99	99%
Charts with Saturn prominent	98	98%
Charts with Uranus prominent	96	96%
Charts with Mercury prominent	95	95%

In the tabulation of the 100 charts by the signs occupied by Sun, Moon and Ascendant, Libra was highest with 32% above the average, Scorpio next highest, and then Leo. The lowest sign was Capricorn, with 36% below average. Taurus was next lowest and then Pisces. No sign was conspicuous enough to be considered a constant. No activity of any special house was significant in the charts of machinists.

To handle tools skillfully and have the energy for the job is mapped by a prominent Mars. Saturn and Uranus thought-cells prominent allow creative ability (Mars) to express easily through mechanical lines. Saturn also coincides with persistence to undergo training, do heavy labor, and use planning and care in doing precise work. A prominent Mercury indicates intelligence to devise better methods, handle new situations or machinery with ease and make repairs. A machinist also needs a retentive memory to handle figures and precise measurements (Saturn-Mercury).

66. Cosmetician

Birth-charts of cosmeticians analyzed	100	100%
Charts with Neptune or Pluto prominent	100	100%
Charts with Mars prominent	99	99%
Charts with Venus prominent	93	93%
Charts with Moon prominent	90	90%
Charts with Pluto prominent	90	90%
Charts with Neptune prominent	88	88%

In a tabulation of these 100 charts by the signs occupied by Sun, Moon and Ascendant, we found Aquarius highest, with Libra and Virgo next. The lowest signs had just about half as many as the highest signs. We were unable to select any particular house whose activity conspicuously contributed to the success of cosmeticians.

The various things done by a cosmetician require mechanical skill, the use of tools, some physical activity and the creative ability to construct something better on the foundation at hand. Thus a prominent Mars is needed.

Artistic taste, refinement of feeling, appreciation of color combinations and form are mapped by a prominent Venus. In addition, the ability to get along with people is mapped by a prominent Moon; the ability to work in cooperation with others by a prominent Pluto, the higher-octave of the Moon; and although not so frequently prominent, Neptune maps the ability to assist the client by dramatizing the best points in the appearance and subduing or making unnoticeable the points of distraction.

67. Birth Out Of Wedlock

Charts analyzed of those born out of wedlock	100	100%
Charts with tenth house afflicted	94	94%
Charts with fourth house afflicted	92	92%
Charts with Venus aspecting Mars	73	73%
Boys' charts with Venus or Mars aspecting Moon, plus girls' charts with Venus or Mars aspecting Sun	92	92%

These births, obtained from the records of a hospital for unfortunate girls, were all timed to the minute.

In the sense that it is something that comes into the life at a definite time after birth when certain progressed aspects form, birth out of a wedlock is not exactly in the same category as the other events discussed in this book. Yet it certainly is an event which, like other important events in the life, tends to condition the individual, even though the conditioning arises gradually over many years subsequent to birth. The aspects mapped at birth significantly portray the conditioning process that usually follows.

Illegitimacy is a restriction imposed by man-made laws. Parents may be unwise to deny social convention, but as the child has no option in selecting the method by which it will be brought into physical existence, the burden of social disapproval should not rightly be attached to it. The handicap so frequently placed upon a child born out of wedlock is not that of organic inferiority, but of man-made social inferiority. This has a marked effect upon the child's conditioning energy. Often exaggerated traits of character are traceable to the unconscious mind's endeavor to compensate for the inferior feeling.

The tenth house rules the reputation and mother. In birth out of wedlock the reputation is affected adversely. 94% of the charts show a badly afflicted tenth. In the cases where strong harmonious aspects also involved the ruler of the tenth, the child was removed to a new environment where the circumstances of its birth were unknown. Even though the child may love its mother, these afflictions map difficulties, distress and turmoil associated with her.

The fourth house, ruling the father and home, was badly afflicted in 92% of the charts. In many cases the place of the often-absent father is taken by others who may or may not be sympathetic. Mother and child may live with relatives to enable mother to work, or the child may be left in the care of strangers, so the privilege of normal home conditions is denied.

Children of parents whose love natures are strong enough to ignore or defy the conventions are apt also to have pronouncedly amorous tendencies. This is the significance of 73 out of the 100 charts analyzed having an aspect between Venus and Mars. It points to society's responsibility to provide opportunity for these children to do creative work, particularly of the artistic type, although these energies may be diverted to other constructive channels indicated by the prominence of some certain planet or planets. Mars and Venus are the two social planets. The Sun rules males in general, and the Moon females. Aspects of Mars and Venus indicate emotional reactions in contact with others. In this study, Venus or Mars commonly made an aspect to the Sun in the girls' charts, and to the Moon in the boys' charts. Due to the intensity of the love natures and to the conditioning that certain attitudes of society may impress on the individual, the opposite sex assumes more than average influence in their lives.

68. WAITER

Birth-charts of waiters or waitresses analyzed	100	100%
Charts with Mars prominent	99	99%
Charts with Moon prominent	96	96%
Charts with Neptune prominent	96	96%
Charts with Pluto prominent	96	96%
Charts with Mercury prominent	91	91%

Sixth House Activity of 100 Waiters

Charts with sixth house active	95	95%
Charts with planet in sixth house	48	48%
Charts with sixth house more discordant than harmonious	40	40%
Charts with sixth house more harmonious than discordant	60	60%
Charts with ruler of sixth actually in tenth or in close aspect to ruler of tenth	72	72%

First House Activity of 100 Waiters

Charts with first house active	94	94%
Charts with planet in first house	62	62%
Charts with first house more discordant than harmonious	42	42%
Charts with first house more harmonious than discordant	57	57%
Charts with first house outstandaing in attractive, affable, magnetic and pleasing personality	88	88%

An analysis of the signs in which the Sun, Moon and Ascendant of the 100 charts are found does not indicate any sign to be a birth-chart constant of waiters. Sun, Moon or Ascendant in Gemini, Leo or Sagittarius seems to map an attraction to this calling more often than other signs do. Aquarius, Taurus and Aries occur least often, having only one-half as many waiters as the other three signs.

Mars maps muscular strength required for the amount of walking and to carry trays of food. Moon maps contact and association with people in general, nutrition, and the foods and beverages served. Neptune maps the courteous, impersonal attitude. Pluto maps groups; usually a waiter serves parties rather than individuals. As Pluto and the Moon are octaves, one or the other is always prominent. Mercury's prominence is associated with manual dexterity, alertness and good temporary memory required. All of these planets are highly prominent in the charts of successful waiters.

Sixth house activity embraces those who are at the beck and call of others. First house activity maps the necessary physical movement as well as the personality, both important for success in this vocation.

69. Telephone Operator

Birth-charts of telephone operators analyzed	100	100%
Charts with Mercury prominent	98	98%
Charts with Uranus prominent	98	98%
Charts with Pluto prominent	96	96%
Charts with Moon prominent	95	95%

Third House Activity of 100 Telephone Operators

Charts with third house active	100	100%
Charts with planet in third house	65	65%
Charts with third house more discordant than harmonious	61	61%
Charts with third house more harmonious than discordant	39	39%

A tabulation of the 100 charts according to the signs occupied by the Sun, Moon and Ascendant gave some variation according to signs, but nothing outstanding enough to be considered a birth-chart constant, or markedly to encourage or discourage anyone from becoming a telephone operator.

A prominent Mercury maps the alertness and quick responsiveness required. A prominent Uranus relates to association with something electrical and various devices which are an essential part of a telephone system. Pluto rules groups of people, and a telephone operator works with and serves others. The significance of the Moon as ruling people in general is obvious. People keep the telephone operator busy. The third house rules calls that are made and conversations taking place by means of the telephone system.

70. Bookkeeper

Birth-charts of bookkeepers analyzed	100	100%
Charts with Saturn prominent	100	100%
Charts with Mercury prominent	97	97%
Charts with Mars prominent	95	95%
Third House Activity of 100 Bookkeepers		
Charts with third house active	94	94%
Charts with planet in third house	55	55%
Charts with third house more discordant than harmonious	55	55%
Charts with third house more harmonious than discordant	41	41%
Eighth House Activity of 100 Bookkeepers		
Charts with eighth house active	91	91%
Charts with planet in eighth house	54	54%
Charts with eighth house more discordant than harmonious	52	52%
Charts with eighth house more harmonious than discordant	43	43%

In the analysis of the 100 charts by the signs occupied by the Sun, Moon and Ascendant, Libra was found to be the highest—probably because Libra dislikes dirty work and tends to neatness and precision—with twice as many as Pisces, which was lowest. Next to highest in number were Gemini and Leo, and next to lowest were Capricorn and Aquarius. No one, however, should be discouraged from becoming a bookkeeper merely because of the sign occupied by Sun, Moon or Ascendant if the planetary prominence and house activity indicate natural aptitude for such work.

We have found handling figures with facility requires a prominent Mercury and Mars; but to be a mathematician requires also a prominent Saturn. To be a successful bookkeeper, while not requiring much mathematical ability, calls upon the Saturn thought-cells because of the dull, monotonous grind of the work. Enough persistence not to give up until the job has been finished right calls upon the experiences incorporated in the Saturn thought-cells. The same Mars mechanical ability enabling a person to handle tools to build things also seems to map ability to use arithmetic as a tool. The prominent Mercury points to alertness and the constant use of intelligence in solving problems.

The nature of the work is probably why so much activity is shown in the third house of bookkeepers. Writing, calculating and thinking are all associated with it. Although not as often as the third, the eighth house is commonly quite active, mapping the work relating to other people's money. We find that to be strongly associated with the things of some one of the twelve departments of life, the individual must have active thought-cells mapped by the house in his birth-chart ruling that department of life.

Although not considered as a constant, Uranus was significant, probably relating to short-cuts and quick methods. Its moderate prominence seems to reenforce the work of the Mercury thought-cells.

71. Stenographer

Birth-charts of stenographers analyzed	100	100%
Charts with Mercury or Uranus prominent	100	100%
Charts with Uranus aspecting Sun, Moon or Mercury	100	100%
Charts with Mercury prominent	97	97%
Charts with Uranus prominent	97	97%
Charts with Mercury aspecting Uranus	68	68%

Third House Activity of 100 Stenographers

Charts with third house active	98	98%
Charts with planet in third house	51	51%
Charts with third house more discordant than harmonious	59	59%
Charts with third house more harmonious than discordant	40	40%

A tabulation of the 100 charts according to the signs occupied by the Sun, Moon and Ascendant gave Cancer highest, then Leo, with Virgo a close third. Taurus was lowest and Pisces next lowest; these two signs showing only about half as many as the high signs. Nevertheless, if the planetary birth-chart constants are shown, there is no valid reason to discourage a person with strong Taurus or Pisces influences from becoming a stenographer.

Quickness in writing down thought, in grasping the ideas of others, and in making time-saving short cuts in placing the thoughts of others on paper is an ability indicated by active Uranus-Mercury thought-cells. Things which come into the life under Uranus stimulus come through a human agency, and stenography is work in which there is close contact with others. It is also the planet of short cuts. Taking shorthand with ease is typical of Mercury-Uranus thought-cells.

Probably the most favorable single aspect for ability as a stenographer is an aspect between Mercury and Uranus. 68% of the charts had this aspect. Next most favorable seems to be an aspect between Uranus and the Moon, with an aspect between Uranus and the Sun coming third. Perhaps we are warranted then in discouraging anyone from taking stenography as a vocation if Uranus is not prominent in his birth-chart.

People who do not have the ruler of the third house powerfully aspected, or at least planets in the third, write comparatively little.

72. Artist

Birth-charts of artists analyzed	100	100%
Charts with Venus prominent	99	99%
Charts with Mars prominent	99	99%
Charts with Neptune prominent	96	96%
Charts with Mercury prominent	94	94%
Charts with Venus aspecting Moon or Mercury	91	91%
Charts with Mars aspecting Moon or Mercury	91	91%
Charts with Venus aspecting Mars	66	66%

A tabulation of the 100 charts according to the signs occupied by the Sun, Moon and Ascendant gave Gemini the highest, with Libra and Sagittarius next highest. The lowest numbers were found in Aries, Taurus and Aquarius, each of which had just two-thirds as many as either Libra or Sagittarius. Even so, if the planetary constants are shown, there is no justification in discouraging a person from becoming an artist simply because his Sun, Moon and Ascendant are in signs which hold fewer artists than some of the others.

The four planets prominent in this report indicate specific abilities. Manual dexterity takes Mercury energy. Dramatic and imaginative presentation takes Neptunian energy. To be able to create there must be ample Mars energy. Appreciation of beauty is mapped by Venus. Thus to create beauty there must be ample Mars and Venus energy. The Moon or Mercury aspecting these planets facilitates the expression in forms of art.

73. Store Clerk

Birth-charts of store clerks analyzed	100	100%
Charts with Saturn prominent	100	100%
Charts with Pluto prominent	99	99%
Charts with Jupiter prominent	98	98%
Charts with Moon prominent	97	97%
Charts with Mercury prominent	96	96%

First House Activity of 100 Store Clerks

Charts with first house active	95	95%
Charts with planet in first house	62	62%
Charts with first house more discordant than harmonious	50	50%
Charts with first house more harmonious than discordant	50	50%

Seventh House Activity of 100 Store Clerks

Charts with seventh house active	94	94%
Charts with planet in seventh house	51	51%
Charts with seventh house more discordant than harmonious	51	51%
Charts with seventh house more harmonious than discordant	49	49%

A tabulation of the 100 charts according to signs occupied by the Sun, Moon and Ascendant gave Libra highest, then Leo, with Pisces third. Taurus and Cancer were lowest with a little more than two-thirds the number to be found in Libra. Next lowest were Aries, Gemini, Virgo and Aquarius; each of these signs containing the same number.

Saturn, prominent in all 100 charts, maps the trade atmosphere in which a store clerk must feel at home, and the required routine that is followed day after day. Jupiter is associated with salesmanship dependent upon good will, an important aspect in the routine of his job. A prominent Moon indicates the people most clerks contact daily. The intelligence mapped by Mercury is necessary to make out bills and keep a record of sales. Pluto was prominent more often than Jupiter, Mercury or the Moon. Pluto rules a store clerk's cooperation; in this case not only with the customers, but with the boss and co-workers as well.

His success depends upon the impression he creates with the public, ruled by the seventh house, which he is compelled to contact daily. To some extent Saturn and Jupiter map it, but his own personality, indicated by the first house thought-cells, assumes prominence. He needs active first house thought-cells for personality and personal activity, and active seventh house thought-cells to attract, for good or ill, the public. People lacking active seventh house thought-cells do not have constant contacts with the public.

74. Electrician

Birth-charts of electricians analyzed	100	100%
Charts with Uranus prominent	100	100%
Charts with Pluto prominent	100	100%
Charts with Mercury prominent	98	98%
Charts with Mars prominent	97	97%
Charts with Mars aspecting Uranus	71	71%

In the analysis of the 100 charts by the signs occupied by the Sun, Moon and Ascendant, Libra and Sagittarius were found to be highest, and Taurus, Gemini and Pisces lowest with not quite two-thirds as many. Next to the highest were Virgo and Scorpio, with Aries and Cancer trailing closely. Aquarius, even though ruled by Uranus, the planet of electricity, had only two more than Pisces and three more than Taurus and Gemini.

Only people with a strong Uranus are attracted to things electrical in nature and associate with them the greatest portion of their daily lives. An electrician is called upon to exercise constant ingenuity in repairing, installing and adjusting a vast assortment of machinery ranging in size from huge generators and dynamos down to minute and delicate instruments. The inventive ability of Uranus, which was prominent in all 100 charts, is called into play.

Pluto relates to hidden forces and the inside of things, ruling inner plane forces and intra-atomic or electromagnetic energies. An electrician need not understand these energies, but his work does require that he become familiar with electromagnetic forces and visualize their activity. Active Pluto thought-cells, prominent in all 100 charts, facilitates his grasp of the subject.

Mercury maps intellectual ability, and Mars mechanical ability. The prominence of the Mars-Uranus aspect favors ingenuity relating to mechanics and building things.

75. Athlete

Birth-charts of athletes analyzed	100	100%
Charts with Mars prominent	99	99%
Charts with Jupiter prominent	98	98%
Charts with Neptune prominent	98	98%
Charts with Mercury prominent	95	95%
Charts with Mars aspecting Jupiter	59	59%

First House Activity of 100 Athletes

Charts with first house active	97	97%
Charts with planet in first house	53	53%
Charts with first house more discordant than harmonious	54	54%
Charts with first house more harmonious than discordant	44	44%

Fifth House Activity of 100 Athletes

Charts with fifth house active	96	96%
Charts with planet in fifth house	55	55%
Charts with fifth house more discordant than harmonious	53	53%
Charts with fifth house more harmonious than discordant	42	42%

A tabulation of the 100 charts according to signs occupied by the Sun, Moon and Ascendant gave Virgo highest, with Sagittarius having almost as many. Cancer was lowest, Capricorn next lowest, and Aries with only one more than Capricorn and two more than Cancer; Cancer having a little more than half as many as Virgo.

The factor in the tabulation not anticipated is the prominence of Neptune, ruling dramatic ability. In other studies Neptune is strongly associated with amusements; in fact, people often take vacations or withdraw temporarily from the more utilitarian pursuits of life under progressed aspects involving Neptune. Most of the charts used here were of those who attained considerable recognition either in the amateur or professional classes. To reach such a goal takes promotional ability as well as the ability to dramatize inborn talents. Another side of Neptune is seen in "fixed contests and games", where the scheming qualities of the confidence-man are involved.

All sports come under the general rulership of Jupiter, indicating "good sportsmanship", the spirit of fair play and the willingness to abide by the rules even if the cost is great. Muscular ability is mapped by Mars, the planet ruling athletes. But as athletics is not merely a matter of brawn, a prominent Mercury maps the fast and sound headwork to size up a situation and make a quick decision.

Neither the number of planets in the first house nor the number in the fifth house can be considered significant. In these same charts 54 show planets in the eighth house and 60, planets in the twelfth. However, the rulers of these two houses are powerfully enough aspected to make the first house (ruling the personal activity and physical body) and the fifth house (ruling sports and amusements) significant.

76. Radio Technician

Birth-charts of radio technicians analyzed	100	100%
Charts with Pluto prominent	100	100%
Charts with Uranus prominent	100	100%
Charts with Mars prominent	100	100%
Charts with Mercury or Moon prominent	100	100%

Ninth House Activity of 100 Radio Technicians

Charts with ninth house active	95	95%
Charts with planet in ninth house	43	43%
Charts with ninth house more discordant than harmonious	56	56%
Charts with ninth house more harmonious than discordant	43	43%

A tabulation of the 100 charts according to sign occupied by Sun, Moon and Ascendant gave Libra highest, with Virgo, Taurus and Gemini next highest. Sagittarius was lowest with two-fifths as many as Libra, and Aries next to lowest with only one more than Sagittarius. Thus, there would be no basis for either encouraging or discouraging any person from becoming a radio technician merely because of the sign Sun, Moon, Ascendant or planets were in.

The electromagnetic waves used in radio are ruled by Pluto. As the technician must associate with and understand these energies, the thought-cells mapped by Pluto were prominent in each of the 100 charts. Because the waves in a radio transmission set are electrically generated, Uranus was also prominent, indicating an attraction to electrical devices, an aptitude for handling electricity, and the ingenuity and inventive ability for making and handling the various intricate parts involved in radio work.

Building and repairing radio apparatus demands the constructive urge and the mechanical ability shown by a prominent Mars. The alert intelligence, cerebral activity and mental ability required in this vocation is mapped by a prominent Mercury or Moon.

Making and repairing sets is not ninth house work. All broadcasting, however, comes under the ninth house. And as instruments constructed are related either to broadcasting or reception of a broadcast, there is at least an indirect association with the ninth house. The technician works with apparatus stimulating ninth house thought-cells. Therefore, it is well for the radio technician to have an active ninth house.

77. Architect

Birth-charts of architects analyzed	100	100%
Charts with Saturn prominent	100	100%
Charts with Mercury prominent	100	100%
Charts with Neptune prominent	99	99%
Charts with Mars prominent	99	99%
Charts with Venus prominent	97	97%
Charts with Uranus prominent	95	95%

Third House Activity of 100 Architects

Charts with third house active	95	95%
Charts with planet in third house	57	57%
Charts with third house more discordant than harmonious	72	72%
Charts with third house more harmonious than discordant	26	26%

Fourth House Activity of 100 Architects

Charts with fourth house active	98	98%
Charts with planet in fourth house	59	59%
Charts with fourth house more discordant than harmonious	73	73%
Charts with fourth house more harmonious than discordant	25	25%

A tabulation of the 100 charts according to signs occupied by Sun, Moon and Ascendant gave Leo highest, Scorpio next; then came Cancer, Virgo and Libra. Pisces and Aries were lowest with only about half as many as Leo.

The careful planning, consideration of precise details, and

painstaking estimates which precede various kinds of structures built upon the land are ruled by Saturn, prominent in all 100 charts. An architect's work is saturnine in nature: plodding and painstaking. The ability to calculate stresses, estimate costs of materials, and ascertain the exact amounts of various required items calls for objective thinking, ruled by Mercury, and facility with mathematics, demanding in addition Mars energy. These two planets are also associated with the skillful use of drafting tools. Neptune maps the imagination necessary to visualize the completed structure. Mars maps creative energy, which to flow persistently into artistic work must be accompanied by a prominent Venus. Visions of beauty, harmony and line must be presented in new designs, and a prominent Uranus maps the actual inventive ability an architect needs.

As his work calls for placing on paper his estimates and drawings, to make this department of life significant enough for attracting such work, he needs active third house thought-cells. The fourth house maps the common thought-cells relating more specifically to the department of life having to do with buildings of all kinds.

78. Dancer

Birth-charts of professional dancers analyzed	100	100%
Charts with Venus prominent	100	100%
Charts with Moon prominent	99	99%
Charts with Neptune prominent	99	99%
Charts with Mars prominent	98	98%
Charts with Mercury prominent	97	97%
Charts with Pluto prominent	92	92%

First House Activity of 100 Professional Dancers

Charts with first house active	97	97%
Charts with planet in first house	62	62%
Charts with first house more discordant than harmonious	61	61%
Charts with first house more harmonious than discordant	39	39%

Fifth House Activity of 100 Professional Dancers

Charts with fifth house active	88	88%
Charts with planet in fifth house	52	52%
Charts with fifth house more discordant than harmonious	50	50%
Charts with fifth house more harmonious than discordant	50	50%

A tabulation of the 100 charts according to the signs occupied by Sun, Moon and Ascendant gave Leo and Virgo highest, with Gemini lowest. Gemini, Aries, Sagittarius and Cancer were low, but had more than two-thirds as many as Leo, Virgo and Aquarius, the three highest.

A prominent Venus maps beauty, harmony and artistic expression; the Moon indicates time, tune and a sense of rhythm; Mars denotes creative energy and muscular strength; Neptune maps the more subtle emotions which bring imagination into play dramatizing inherent talent; Mercury energies are needed to thoroughly memorize long and intricate routines; and professional dancing is often done in groups, requiring an active Pluto.

The vigorous personality and sustained physical activity of the successful dancer is mapped by a strong first house. As professional dancing is work which has to do with entertainment, fifth house associations during the time the profession is followed are important. Therefore, to attract these associations fifth house thought-cells should be active.

79. Chemist

Birth-charts of chemists analyzed	100	100%
Charts with Pluto prominent	100	100%
Charts with Saturn prominent	100	100%
Charts with Neptune prominent	99	99%
Charts with Uranus prominent	98	98%
Charts with Mars prominent	95	95%
Charts with Mercury prominent	93	93%

Sixth House Activity of 100 Chemists

Charts with sixth house active	93	93%
Charts with planet in sixth house	48	48%
Charts with sixth house more discordant than harmonious	58	58%
Charts with sixth house more harmonious than discordant	42	42%

A tabulation of the 100 charts according to signs occupied by Sun, Moon and Ascendant gave Leo highest, with Virgo and Libra having almost as many. Taurus was lowest with a little over half as many as Leo; and Gemini and Cancer were next lowest.

Chemistry relates to changing inside factors of compounds, of manipulating groups (Pluto) of atoms so they separate from, or combine with, other atoms to form other molecules. To understand how these invisible forces within the molecules operate and to handle them successfully seems to require active Pluto thought-cells. It appears that Pluto is the chief ruler of chemistry and therefore of chemists. A chemist cannot afford to hurry his work, and no guesswork can be tolerated. A man of system and order, he must be deliberate and persistent. Thus, a prominent Saturn is a requisite to this vocation.

Gases, chemicals and poisons in general, ruled by Neptune, form part of the surroundings in which a chemist spends most of his time. Unless Neptune is active, he may not be attracted to the environment. Research comes under Uranus as well as complicated tools and the apparatus he is forced to invent himself.

Construction of things in his laboratory requires intelligent and mechanical ability, mapped by Mars and Mercury. A keen mind, hard study and manufacturing skill are needed to follow successfully the vocation of a chemist.

Chemicals, and drugs in particular, come under the rule of the sixth house. Therefore, unless the individual has active sixth house thought-cells, it is unlikely he will be sufficiently attracted into association with these things to become successful as a chemist.

80. Dentist

Birth-charts of dentists analyzed	100	100%
Charts with Jupiter prominent	100	100%
Charts with Mercury prominent	100	100%
Charts with Saturn prominent	99	99%
Charts with Mars prominent	98	98%
Charts with Venus prominent	98	98%

Sixth House Activity of 100 Dentists

Charts with sixth house active	97	97%
Charts with planet in sixth house	44	44%
Charts with sixth house more discordant than harmonious	58	58%
Charts with sixth house more harmonious than discordant	37	37%

A tabulation of the 100 charts according to the signs occupied by Sun, Moon and Ascendant gave Aries and Virgo highest, and Sagittarius lowest with a little more than half as many. Next to highest was Leo, and next to lowest were Gemini and Cancer.

Active Jupiter thought-cells stimulate the individual to take up a profession, giving warmth and geniality which cause people to like him and come to him for service. Optimism encourages confidence in his skill. A dentist often keeps people entertained while he drills their teeth and hammers repair material into cavities. Many dentists divert the patient with interesting conversation while doing the necessary work. All these things stem from active Jupiter thought-cells.

Study, alertness and a high degree of dexterity — fine manipulation with the hands — required of a successful dentist depend upon the Mercury thought-cells. His chief job is to save teeth, ruled by Saturn, and make them last as long as possible. He exercises great ingenuity saving and not removing them. Dentistry is mainly a Saturn occupation.

It is also work requiring mechanical skill, and the building of something, and therefore needs Mars thought-cells active. Neither Mars nor Saturn, both being harsh, conduce to sensitivity. A person with these two planets prominent does not mind too greatly the sufferings of others, and a dentist should not be hypersensitive. To keep his patients coming back, he needs a certain degree of sociability, mapped by Venus, whose thought-cells are stimulated when he is called upon to fit teeth artistically and attractively.

The ailments of people, including difficulties they have with their teeth, come under the rule of the sixth house. The dentist depends upon people with sick teeth for his livelihood. And unless he has active sixth house thought-cells, he is not sufficiently attracted to ailing people to get their patronage. Thus, a dentist should have an active sixth house.

General Reports

81. Nodes, Part of Fortune, Fixed Stars

It is one of the cardinal doctrines of science, applicable to every line of research, that so long as a consideration can be adequately explained by factors already recognized, no new factor should be introduced into its explanation. Both in delineating the birth-chart and its progression, we have held tenaciously to this principle, upon which material science has so successfully been able to build its systems.

We are sometimes asked why we neglect the Moon's Nodes, the Part of Fortune, and the Fixed Stars in the birth-chart. We do so because, up to date, we have not found any condition in a person's life which could not satisfactorily be explained by the planetary positions without recourse to these other factors. We do not say that these positions have no value; merely that up to the present we have found no need to use them in explaining the conditions and events in people's lives.

The part of fortune is a point on the ecliptic as far removed from the Ascendant by longitude as the Moon is removed from the Sun by longitude. The Moon's nodes are the dragon's head and tail.

Nodes refer to the point where the orbital paths of two heavenly bodies intersect. As other planets do not move in the same plane as the earth's orbit, each planetary orbit must cut the orbit of the earth at two points. Therefore, the Moon and each of the planets have an ascending node and a descending node. If nodes are to be used as sensitive points, logic would indicate that the nodes of all planets, as well as those of the Moon, should be used.

Although many astrologers who have attained outstanding reputations for accuracy in the practice of horary astrology have started by using the part of fortune and the Moon's nodes, it became apparent to them as they gained greater proficiency that these symbols in a chart were a hindrance to the reliability of their work.

From the standpoint of non-astrological divination, the more numerous the symbols present — from which Extra-Sensory Perception may take its choice in giving correct impressions to objective consciousness — the better. But from the standpoint of the Hermetic System of Astrology, research compels us to employ only those factors which actually equal or add accessory energy to the thought-cells of the unconscious mind.

It is our view that for astrology to be a science, equivalents must be established within the unconscious mind of the human being for every factor in the birth-chart. To the best of our knowledge, no one has established such equivalents for the Moon's nodes, part of fortune or fixed stars.

A vast amount of statistical work in natal astrology and checking predictions in mundane astrology impels us to recommend that fixed stars, the part of fortune and the Moon's nodes be omitted from all astrological charts.

82. Diurnal Revolutions

A diurnal revolution is a chart erected for the moment the Ascendant on the given day reaches the sign, degree and minute of the zodiac occupied by the birth-chart Ascendant. The chart should be erected for the longitude and latitude occupied by the person at the time. Our research did not find diurnal revolutions reliable in indicating what will happen during the following day.

83. Lunar Revolutions

The transiting Moon making the conjunction with its birth-chart place is one type of lunar revolution. The other, and more creative period, is when the transiting Moon forms the conjunction with the natal Sun. In either case, the chart should be erected for the longitude and latitude occupied by the person at the time.

Astrological research has not found lunar revolutions reliable in indicating what will transpire during the following month. Some astrologers allot great importance to the birth-chart house occupied by the New Moon, but astrological research found this house of no more importance than ordinary transit positions. After conducting a great deal of research on this matter it was found that an eclipse of the Sun or Moon falling on a birth-chart luminary or other birth-chart position is of no more significance than a heavy transit.

84. Solar Revolutions

A solar revolution is a chart erected for the moment the transiting Sun returns to the same sign, degree and minute of the zodiac it occupies in the birth-chart. The chart should be erected for the longitude and latitude occupied by the person at that time.

Some astrologers allot great importance to this type chart, but astrological research has not found solar revolutions reliable in indicating what will transpire during the following year. However, the time of the Sun's transit over its birth-chart position seems to correspond with a valuable creative period that can be used to great advantage.

85. Minor House Cusps

Events occur in people's lives which can only be explained adequately through the progressed aspects of the Ascendant and Midheaven. They are progressed as if they were planets. But in the birth-chart, no aspects to the cusps of other houses are calculated and we do not progress planets to aspects of the cusps of any other houses, nor progress the cusps of other houses in any way in the Hermetic System of Astrology; because so far very extensive astrological research work reveals nothing that cannot be explained without such aspects and progressions. The term minor house cusps refers to the ten other cusps of the birth-chart aside from the first (Ascendant) and the tenth (Midheaven).

86. Animals and Astrology

Our research found that stimulation mapped by progressed aspects affects animals in the same manner it affects men, due allowance being made for the normal level of the animal. It is not to be expected that an insect with a few-days life span will respond other than to progressed positions within those few days. Just how they respond could only be judged by timing their births and calculating subsequent aspects. But we have had ample opportunity to observe that cats, dogs, birds and horses, on their level, react under progressed aspects as do human beings.

In a series of timed and progressed birth-charts of dogs, for instance, a harmonious progressed aspect to the tenth house indicated with close certainty the dog would win a show prize. The same stimulation in a politician's chart predisposes to winning an election. Another series of charts on race horses revealed when the horses were subject to accident, to obstacles of a given race, and if the horse was worth training in the first place.

We have charts of birds and dogs with the most important events in their lives listed and the progressed aspects noted. The same constants are present as appear in the coincident events with a person's chart. Furthermore, in every case, each major progressed aspect was reenforced by a minor progressed aspect and triggered by a transit aspect to one of the involved terminals, just as in the case of a human birth-chart.

87. Mundane Astrology

The Church of Light Astrological Research Department diligently investigated the birth-charts and progressed aspects of all countries and towns for which accurate birth-data was available. Research indicates that while countries and sections of countries respond more pronouncedly to Cycles and current astronomical phenomena than individuals do, they — as well as cities and lesser corporations—react much as people do to the stimulation mapped by major, minor and transit progressions.

As a nation or a section of the country responds markedly to Cycles and current astronomical phenomena, they should receive special consideration even when the birth-chart is known. But if reliance should have to be placed on only one influence, far greater precision can be gained through the use of major, minor and transit progressions of the birth-chart than through the use of Cycles and current astrological positions.

It was found that in predicting events for a nation, city, town or corporation, nothing else gives the details and precision that can be gained by the use of a correct birth-chart and progressed aspects.

Because of the difficulty in getting correct birth data, Cycles and Major Conjunction Charts gave much information as to what would happen at a specified time within a given area. However, these charts did not clearly designate the city or locality chiefly affected by the event. They did not point out the one specific enterprise, among many of a similar nature, which felt the full weight of the stimulus. Such details can only be revealed by progressions calculated to a correctly timed birth-chart.

Other aspects of the more slowly moving planets have a noticeable affect upon the affairs of cities, nations and the world. And a chart erected for the city or capital of the country for the exact time the aspect is perfect will give much information as to the department of life chiefly affected by the aspect.

Such aspects in the sky map some influence during the time they are forming and during the time they are separating over an orb for the same number of degrees allowed for the same birth-chart aspect. In this they are quite different than the aspects made by progression in a Cycle Chart which—with the exception of the swiftly moving Moon—map activity only during the time the aspect is within one degree of perfect.

88. Group Leaders and Mundane Astrology

All people are not equally receptive to vibrations being broadcast from the sky; no more so than are all nations. Some individuals within a group may tune in scarcely at all on a given planetary vibration, while a few members may pick up its energies in tremendous volume. These members responding markedly to a planetary vibration largely determine the effect of the particular energy upon the group.

Those having the planet prominent in their birth-charts are more responsive to its vibrations than are those who do not have it prominent. That is, they have thought-cells within their unconscious minds acting as natural receiving sets for this energy.

But to pick up the planetary energy, radio fashion, in most volume, there must be at the time a progressed aspect to that planet in the birth-chart. Such a progressed aspect forms an aerial which readily picks up the energies of the two planets involved. This energy is conducted immediately to the thought-cells mapped in the birth-chart by the planets. And these thought-cells, acquiring so much new energy of their own type, powerfully influence the individual's thinking, his behavior and the events attracted into his life at that time.

Astrological research afforded ample opportunity to work out the progressed aspects in the charts of these individuals whose activities were chiefly responsible for an effect upon a nation or a city coincident with heavy aspects in the sky at the time. Such individuals had the aspecting planets prominent in their birth-charts, and at the time there was a major progressed aspect to one or both of the birth-chart planets.

It was found that the fortune of a nation usually coincided with the progressed aspects of its leader. A study of the United States Presidents during whose terms of office there was either a war or a nation-wide financial depression since 1800 demonstrates the point. (These birth-charts with declinations and case histories appear in the book "Horoscopes of the U.S. Presidents", by Doris Chase Doane.)

Abraham Lincoln was elected president in 1860 under progressed Jupiter trine natal Saturn, a long-time favorable aspect coincident with his whole rise to political power. At birth Mercury was in close sesqui-square aspect with Mars, and when the Civil War reached its peak, his progressed Mercury formed the inconjunct aspect with progressed Mars.

Ulysses S. Grant was elected president the second time in 1872 under progressed Venus trine natal Uranus. His natal chart shows Mars sesqui-square Uranus, and the year following his second election progressed Jupiter was closely square natal Mars and sesqui-square natal Jupiter. On September 18, 1873, a financial panic started a severe depression lasting through his term, and all that time the Jupiter affliction was operative.

Grover Cleveland was elected the second time in 1892 while progressed Sun applied to a square of progressed Mars. Near the time when this aspect reached its peak, only four months after his inauguration, a nation-wide commercial panic started.

William McKinley was elected under progressed Venus conjunction natal Jupiter, and both planets were moving to square progressed Mars during the Spanish-American War in 1898. When he was reelected in 1900, progressed Sun was sextile natal Jupiter, but in 1901 when progressed Mercury squared natal Mars there was a financial panic in Wall Street.

Woodrow Wilson was elected the second time in 1916 under progressed Sun sextile natal Sun and trine natal Saturn, but at the time, his progressed Saturn opposed natal Sun and progressed Uranus tied into the progressed Saturn with a semi-square aspect. The United States entered World War I.

Herbert Hoover was elected in 1928 under progressed Sun conjunction progressed Jupiter and trine natal Saturn. But in November of 1929 when his progressed Saturn opposed both natal Moon and natal Mars, and progressed Venus opposed his natal Ascendant, the worst period of economic depression started in this country.

Franklin Delano Roosevelt was elected in 1932, and the following year launched the New Deal under progressed Ascendant trine natal Mercury and trine natal Mars. Then in 1940, when progressed Mars was sextile progressed Saturn, he was elected for the third time. By December 7, 1941, the day of Pearl Harbor and the formal entry of the United States into World War II, his progressed Sun squared progressed Mars at the same time progressed Mercury opposed his natal Uranus.

Harry S. Truman inherited the presidency just as he inherited the war. In 1948, his natal Sun was in trine aspect with progressed Mars. At the time his discordant major progressions were passing out of orb to make way for three sextiles and a semi-sextile, which pictured a quick ending to the war, high wages, full employment and no financial depression.

Not only is this trend seen in the charts of presidents and national leaders, but it holds for any type of group leader whether a great religious potentate leading millions of people, or the chairman of a six-member poetry group. Progressed aspects to the leader's birth-chart picture the fortune of his group in a broad manner.

89. Significance of Eclipses

The Church of Light Astrological Research Department collected a large number of instances in which either a Solar Eclipse or a Lunar Eclipse took place in the same zodiacal degree occupied by the Sun and other planets in people's birth-charts. In none of these did the eclipse coincide with events that were not clearly and fully accounted for by the progressed aspects at the time.

We have not been able to verify the doctrine that the power of an eclipse persists, even if a long eclipse, over a period of years. We have, however, checked the influence of every Solar Eclipse since 1884 that was visible in a part of the world fairly well populated. This survey lead to positive results. It was found that if a Solar Eclipse occurs in a region where there is considerable population, within a few months before, or much more likely within a few months after the eclipse, there is a disaster in the region where the eclipse is visible.

While the disaster tends to be near the central path of the eclipse, it may be anywhere the eclipse is even partially visible.

It seems likely that the exact place of the disaster is determined by the progressed aspects in the birth-charts of cities and regions, and the progressed aspects in the Cycle charts erected for those places. Because the correct birth-charts of so few cities are known, it is difficult to determine where within the visible area of a Solar Eclipse the disaster will take place. However, the New Moon chart often revealed the nature of the disaster.

From the standpoint of Natal Astrology research points out that an eclipse of the Sun or Moon falling on a birth-chart luminary or other birth-chart position is of no more significance than a heavy transit.

90. Upper-Octave Planets

In the divine alphabet of the language of the stars, there are 22 letters. The twelve zodiacal signs correspond to the consonants, and the ten planets correspond to the vowels. This may seem strange to those conversant with modern opinion that the ancients, having no optical instruments, could know only planets visible to the unaided eye. It must seem they could not have known Uranus, Neptune and Pluto.

Yet the attributes associated with these three upper-octave planets are presented clearly and unmistakably in ancient mythology. After painstaking study of Neptune, Uranus and Pluto, the modern astrologer notes their attributes, which, experimentally demonstrated, are in detail the same assigned to the mythological characters Uranus, Neptune and Pluto. This is more than mere coincidence.

There are those who, even today, eliminate these three "spiritual planets" from a horoscope, saying the masses have not evolved to a response-level enabling them to receive stimulation associated with the wayward planets. In this we cannot agree. Our research revealed that not only do individuals react to stimulus mapped by the upper-octave planets today, but people also reacted before their discovery. Research also shows that new planets usher in new periods in world affairs. These trends depend upon individuals within a group, who—of necessity, explained by the Law of Association—capture and release astral energies of the type mapped by Uranus, Neptune and Pluto, working to bring about the new period in world affairs.

It may be that beyond the orbit of Pluto, which marks the present day frontier, lie other yet to be discovered orbs. When they are discovered, provided they exist, our observation of world affairs coincident with the findings of Uranus, Neptune and Pluto suggests that each will usher in a new era in world progress.

Uranus, planet of invention and independence, discovered in 1781, may rightly be said to have ushered in the machine Period and the Period of republics. At that time the looms of England rapidly transformed that country from an agricultural into an industrial nation. Engines of all kinds came into use—one invention following another. Still later electricity added its power to this Period of manufacture. In 1782, the Peace of Versailles and Paris granted independent existence to the United States, setting a precedent to be followed by most countries of the New World. The Republican form of government came to be a dominant factor in the Western Hemisphere.

Discovered in 1846, Neptune ushered in the Period of oil and gas, preparing the way for our present industrial system and means of locomotion. In 1848 modern spiritualism, a new religious conception, had its birth, giving the world scientific proof that the soul survives physical death. Before the Fox Sisters and their rappings in that year, similar phenomena was labeled the work of the devil and witchcraft. Aside from revolutions in Europe and the discovery of gold in California, this Period of oil coincided with new methods of financing business—the pooling of individual resources. Instead of individual enterprise and partnerships—with limited ability to raise capital—corporations composed of share holders dominated the business and industrial world.

The Pluto Period commenced with the official announcement of its discovery on March 13, 1930, only a few months after the 1929 collapse of the stock market starting the greatest financial depression the world ever experienced. Kidnaping developed into a lucrative crime, and large scale racketeering, bred of prohibition days, reached its peak. The New Deal popped into politics, the sit-down strike held up industry, and the world became divided into two camps—one dominated by militant and predatory dictators, the other a defensive cooperation of democracies.

A study of historical events and the coincident astrological configurations reveals a close association. Many outstanding individuals were active in bringing the foregoing changes to bear upon the particular period in which each change manifested. The birth-charts with progressions of the individuals reveal active thought-cells mapped by the higher-octave planets.

It is true that after a particular Period ruled by a planet began, environmental conditions were more abundant in which the planetary energy could express, and it would seem that the attribute ascribed it assumed more importance in these individuals' lives after the discovery. But our astrological research on the charts of people who lived before the Periods of each of the upper-octave planets started, shows conclusively that despite astrological ignorance of these planets, they were even then mapping events that profoundly affected people's lives through such conditions in the environment then at hand.

91. Transits and Hermetic Astrology

The most controversial of all astrological subjects relates to how much dependence can be placed on transit aspects. If an orb sufficiently wide is allowed, a transit aspect can be found that appears to account for every event of life. But when so wide an orb is used, some such transit aspect is present at all times, and there is nothing which clearly shows which of these numerous transit aspects indicates an event, and which does not. Under such circumstances the astrologer who has active extra-sensory perception is able to pick the one which will coincide with an event. But those who rely exclusively on the aspect cannot do so; for while events do often coincide with transit aspects to birth-chart positions, innumerable transit aspects to birth-chart positions occur which do not coincide with any significant event.

The Church of Light has never maintained that transit aspects have no influence, but it has taught that their importance has been greatly over-estimated by some. Furthermore, the Hermetic System of Astrology pays no attention whatever to the part of fortune, Moon's nodes, progressed minor house cusps, or the Arabic or so-called sensitive points, because statistical analysis of thousands of charts progressed to the time of an event indicates there is no need to consider them.

Both in the birth-chart and by progression this system of astrology confines its attention exclusively to the positions of the ten planets, the Midheaven and Ascendant. For judging by natal astrology events probable in an individual's life in order to suggest precautionary actions, it pays no attention to Cycles or other such charts; having discarded their use after careful statistical analysis proved them unreliable. It was revealed that in addition to the birth-chart the only reliable factors were major, minor and transit progressions.

After a vast amount of statistical research on the matter we find the following rule fully justified:

Rule: Each reenforced major progressed constant of an event is always released by a transit aspect heavier than from the Moon to one of its four terminals at the time the event occurs. And an independent minor progressed aspect is always released by a transit aspect to one of the birth-chart or major progressed terminals influenced by the minor progressed aspect at the time of the event.

Transit progressed aspects have two distinct influences. The psychokinetic power of the thought-cells receiving new energy through transit progressed aspects enables them to attract into the life inconsequential events. Inconsequential events coincide with characteristic transit progressed aspects. But in addition to this independent influence, if the transit progressed aspect is to one of the terminals of a major progressed aspect, it has a trigger effect, tending to release either the minor progressed aspect or the major progressed aspect, or both.

Ample statistical data indicates that transit aspects have very little power in themselves, but that they exert a trigger effect which tends to release the power of reenforced major progressed aspects, and also the power of independent minor progressed aspects.

It will now be apparent why some think transits are so powerful. At the time events happen there is always a transit aspect within one degree of perfect involving the planets mapping important events. However, those who ignore major progressed aspects, on the average miss one-half of such significant transit aspects; for the transit aspect to the major progressed position of a planet is as powerful as the same transit aspect to the birth-chart position of the same planet.

The main point, however, is that research found that no transit aspect coincides with an important event in an individual's life unless at the *same time* there is a major progressed aspect involving the aspected planet which is reenforced by a minor progressed aspect.

Summary: Statistical research convinces us that for major events primary reliance should be placed only on the major constants of the event. Minor events take place only during the periods indicated by minor progressed aspects, and major events take place only during the peaks of power indicated by minor progressed aspects which reenforce *all* the major progressed constants of the event. And the actual event takes place only when this reenforced power is released by the trigger effect of a transit aspect to one of the terminals of *all* the major progressed constants of the event.

92. Significance of New Stars

The wise men of the east looked to the heavens to apprise them of important events to come. Unusual phenomena in the sky, to them, portended unusual events which would happen on earth. According to the rules they left, the appearance of a new star signified the commencement of a new condition in the world, which would have far reaching effects upon the affairs of men.

The phase of human activity in which a turning point had been reached, and henceforth a new condition would manifest, was indicated by the constellation in which the new star appeared. As both the pictures of the constellations which they used and the stories about them are still accessible, our Research Department decided to apply them precisely as these ancients used them in their work.

We tested the same sules by studying world occurrences which immediately followed similar phenomena recorded in the past. Astronomers estimate that during the past 2,000 years there have been about 30 new stars of sufficient brightness to be seen without a telescope. Only 12 of them, however, in addition to the Star of Bethlehem—the location of which is unrecorded—have been of third magnitude or brighter. It was not a difficult task, therefore, to apply the ancient rules to all the conspicuous novae (called new stars, but probably old stars which have exploded) that have been recorded since before the birth of the Nazarene. The tie-up of historical events with each new discovery is presented in the book "Mundane Astrology", by C. C. Zain, covering a period from early times down to the discovery, in 1934, of Nova Herculis in the constellation of Hercules.

Hercules, as pictured in the sky, is represented on one knee, with his other foot crushing the head of a dragon that winds its slimy coils of graft and corruption around the northern axis of the world. He holds aloft the fruits of his toil. Other objects no less significant are in his hands; but according to the ancient rules, because the New Star is in the vicinity of the dragon-crushing foot, the stamping out of unfair dealing is the most striking feature foreshown. Subsequent events point to the struggle for power and the rewards of labor, activities having a tremendous influence over the political and business life of the world.

After painstaking study of the New Stars discovered in the past two thousand years and the world trends which developed after each discovery, our research department found that the appearance of New Stars, even though not part of our solar system, signify some new trend in the world, pictured in universal language by the constellation in which it appears.

93. Significance of Comets

An opinion quite widely accepted is that comets, when they enter the zodiac, bring with them new conditions affecting the affairs of men. The appearance of the more important ones in the past have always coincided with unusual events upon the earth. The old rule was that the influence would be felt chiefly in the country ruled by the sign in which the comet was first visible. Thus just preceding the great debacle of the Russian armies in World War I and the revolution that followed, a comet—which later developed to important size—was discovered by means of a telescope in the sign Aquarius, ruling Russia. Astrologers the world over began to predict that startling things would happen in that country—predictions that were fully verified.

As comets actually belong to our solar system, which New Stars do not, comets should be referred not to their place among the constellations, but to their place in relation to the zodiac. Perhaps now that they may be discovered by telescopes long before they enter the zodiac, the signs in which they appear after thus entering the zodiac indicate more precisely the regions of earth affected.

For example, Peltier's Comet, on August 3, 1936, entered the Zodiac at about 28 degrees Aquarius, to leave again August 6, 1936, at about 14½ degrees Aquarius. The following year and a half witnessed blood purges in Russia, ruled by Aquarius, when most of the important men who assisted in establishing the Soviet Union were executed on charges of trying to overthrow the government.

Comets vary in size, in shape, in brilliancy, and even in color. Some of them, it is true, are periodic, but even these do not have the same appearance on successive returns; and may go so far as to be beyond visibility in even the strongest telescope. Others come into our solar system from the spaces without, bringing their own astrological vibrations, and after arcing about our Sun, pass on into space, never again to return. Having no previous acquaintance with such celestial visitors, their influence can not be known from earlier observations.

Some comets in the past have coincided with pestilence, with great wars, some with disasters, some with revolutions, some with great constructive enterprises, and some with the birth of industrious persons. The general rule has been that the shape and apearance of the comet signified the nature of its influence. If it looked red and angry, it signified disaster. If it looked like a sword it meant war. If it had a pleasing appearance, it heralded some great constructive enterprise.

As the matter stands there is need for much research as to just what may be expected from a given comet. The only reason the Church of Light Astrological Research Department has not already contributed to this knowledge is that the literature in which descriptions of the old comets are given seldom specifies where they appeared in the zodiac or in the celestial sphere.

94. Rectifying the Horoscope

A birth-chart is a map of the thought-energies within the astral body of man, and the progressed aspects correspond to the time and nature of the innerplane weather added by planetary energies. In natal astrology, then, the chart becomes a means of estimating the various forces within man's finer body.

One of the objects of natal astrology is to indicate not merely how the energies reaching the astral body of man may accurately be estimated as to volume and their natural trend, but also to indicate the methods by which they can be controlled, diverted and employed to do the kind of work man most desires of them. For accurate estimation of their power and for sound instructions by which they can be manipulated to perform a more constructive purpose, the birth-chart (forming the basis upon which these matters are judged) must be accurate.

Unfortunately, some people do not know the hour of their birth. Many methods and glowing promises of rectification are presented on all sides, and some astrologers even boast that they never judge a chart, even when the time of day is given, unless they first rectify it.

In our research department we have found it better always to use the time of day given. The chart may be a few degrees off, but the error introduced is usually smaller than the error commonly occasioned in trying to make the chart fit some theory Theories should be made to fit facts, and the observed time of birth is the fact in this case that all theories should be made to fit.

As a result of studying hundreds of charts, our research shows that the best means of rectifying a horoscope is by checking events in the life against the Birth-chart Constants and Progressed Constants listed in the foregoing research reports. Methodology for rectification embraces seven steps.

1. *Case History.* The more that is known about the native the better. Because an individual is more than that pictured by his chart of birth, as wide a knowledge as possible of the conditioning he has experienced since birth should be acquired. The type of fortune he has had in each department of life, how he feels and reacts to things in each department of life, and what he attracts to himself from environment mapped by each department of life is significant, as well as the dates (as near as possible) when events occurred relating to each department of life. The Church of Light Astrological Research Department recently issued a revised questionnaire which is helpful in compiling such a case history. A copy follows:

GIVE ASTROLOGY A HELPING HAND

Help yourself by helping others. The more astrological research we undertake, the more facts we can disseminate to insure accurate interpretation of horoscopes in order that precautionary actions can be taken. Will you assist us in this great work by answering this questionnaire? We want all the case histories we can get. If you can get others to cooperate in this venture, we will be pleased to send more forms on request.

INSTRUCTIONS: Even though you cannot answer all the questions, send any information that you can. Only cases where birth hour is known are useful to us. When "he" appears, it refers to the native regardless of sex. If spaces after questions are not large enough, kindly use another paper for answers, copying the number and letter of question to be answered. *Please give DATES, TIMES and PLACES as near as possible whenever you can.*

RESEARCH QUESTIONNAIRE

Submitted by .... Date ....
(name and address please)

Name of native .... Sex ....
(kept confidential by us)

Birth month, day and year ....Birth hour a.m. or p.m...........
(kind of clock time)

Place of birth ....
(Town County, or county seat State Country)

If rural community, name nearest large town..............................

Its distance from birth place..........................miles.

Personal nationality ....How many generations..........

Nationality of ancestors ....

PHYSICAL DESCRIPTION: Include photo or snapshot if possible.
Hair
Height
Teeth
Eyes
Weight
Nails
Skin
Body Characteristics
Scars, birthmarks, etc.
Describe walk
Actions: fast/slow:
1a—Is vitality strong?
b—What can be said about personality?
c—Was early childhood happy?

d—Give birth data (date, hour and place) of grandmother on father's side:
and birth data of grandfather on mother's side:

2a—Has he had success in money matters, or financial reverses?
b—What kind of possessions does he gather?
c—How does he care for his possessions?
d—What philosophical importance does he attach to money?

3a—Is he studious?
d—Difficulty in learning?
c—Describe interest and ability in education.
d—Graduate High School, College, or other school?
e—How does he get along with relatives and neighbors?
f—Does he have a drivers' license? Ever get a ticket? When and why?
g—Does he travel much? Name places, dates, conditions under which trips were taken and events which occurred while enroute.
h—Does he have a pilot's (airplane) license? Any legal trouble in connection with it?

4a—Does he own his own home? What other property?
b—Describe home life. Has it been harmonious? Give general surroundings and atmosphere of home briefly.
c—Describe father and relationship to him (physical, mental and emotional).
d—Give as complete birth data (date, hour, place) of father as possible.
e—If native is elderly, describe present life conditions.

5a—What success in love?
b—What have been the great emotional periods in life? Give dates.
c—How many children? Give complete birth data on children and grandchildren.
d—How does he get along with his children? Other children?

e—What is his favorite color? If more than one, give order of preference. Does he dislike any color? If so, was there something discordant in the past associated with that color? Has color preference changed since childhood?

f—What are his favorite hobbies?

g—Describe interest and abilities in speculation, gambling and entertaining.

6a—What are his favorite foods?

b—Do any foods disagree with him?
Food Poisoning? Dates.

c—Does he like pets? What kind?

d—What kind of illnesses has he had? How serious? When? (Please use extra sheet and give as many details as possible.)

e—Accidents? All types, dates and conditions.

f—How does he go at his work? Get along with co-workers?

g—Does he habitually finish the things he starts?

h—Has he served in the armed forces—dates of entry and discharge?

i—Has he been involved in labor strikes? When?

7a—Single, married, widow or widower?

b—If married, is marriage happy? Describe relationship.

c—Mate's complete birth data: date, hour, place of birth.

d—Data of marriage: date, hour, place.

e—If in business partnership, describe relationship and partner.

f—Business partner's complete birth data:

g—If divorced: give dates, conditions and reasons, etc.

h—Has he suffered from partners, open enemies, or law suits? Dates, why?

i—What is relationship with grandmother on mother's side, give her birth data:
Relationship with grandfather on father's side?
Give his birth data:

8a—Has death or legacy played an important part in his life?
b—Deaths: dates in family, friends, pets, etc.
c—Inheritances: What types? Dates?
d—Have debts played an important role in his life?
e—Has he been approached many times to lend money? Has he tried to borrow money often? Explain.
f—Ever do any banking? Or handle money for others?
g—Has he had any psychic experiences? Describe and give dates.
h—How has insurance affected his life? Dates.

9a—What is the nature of his religion?
b—In regards to philosophy, does he practice what he preaches?
c—Has he appeared in court? If so, explain, give outcome and dates.
d—Has he been published: books, magazines, etc. Type and dates.

10a—What is his occupation? Is he successful?
b—What other occupations has he followed? Success?
Any jobs with emphasis upon honor rather than wages? If possible, give dates when he was hired, fired, discharged or quit. Describe conditions and how job was obtained.
c—What is his position in the community?
d—What honors has he received? Type and dates.
e—Does he enter into community activity?
f—Mother's complete birth data (date, time, place) and his relationship to her.
g—Describe relationship with superiors.

11a—What kind of friends does he attract?
b—Has he many friends? Do they help him? Explain, give dates.

c—What kind of desires does he have?
Does he attain them?

d—If given three wishes, what would he wish for?

12a—Has he had many disappointments and sorrows? Explain, dates.

b—Has he suffered from secret enemies? How, dates.

c—Has he been institutionalized—hospital, asylum, jail, sanitarium, etc.—explain and give dates.

d—Has he ever worked in an institution? Name type of work, dates.

e—Has he ever been connected in any other capacity with an institution?

f—Has he ever donated blood? How many times? Dates. Affect on system.

COMMENTS (Please add any other information which might be of interest in this astrological project.)

Please mail to: ASTROLOGICAL RESEARCH DEPT., The Church of Light, P. O. Box 1525, Los Angeles 53, Calif.

Thank You

2. *Rising Sign.* Occasionally the rising sign marks the person quite clearly, but in most cases a combined impress of other influences make difficult the judgment of the rising sign. Unless there seems to be a clear-cut indication, it is best not to guess at the Ascendant.

3. *Trial Chart.* Using the chart indicated by the rising sign, if one was obtained, or a Natural Chart (no-degrees Aries on the Ascendant, Taurus on the second house cusp, etc.) relate the information in the case history to each of the 12 houses in order to establish as closely as possible the general significance of the planets in the houses. The trial chart is turned until it reaches a place of reference to these positions.

4. *Check Ephemeris for Progressions.* From the day of birth in the ephemeris, follow down the page and note the chief aspects that form between the progressed planets, and between the progressed planets and birth-chart planets. The outstand-

ing major progressed aspects will indicate an event which happened in a certain year. But this alone will not reveal exact dates within that year, details of the event or even the department of life affected.

It is because events can be described in general terms and with approximate dates, without house positions, that it is absolutely essential in rectifying a horoscope to carefully distinguish between (1) the general influence commonly associated with a planet and (2) the department of life involved.

Check the Constants for various events with the progressions appearing in the ephemeris for the corresponding years. Then turn the trial chart and manipulate it until it reaches a place of reference to the house positions of the planets, and the planetary rulership of the houses fitting the known facts of life. After a few events seem to click, each house in turn should be given due consideration as to the events that have occurred in the department of life signified, to make sure if the planet in it, or ruling its cusp, indicates such fortune. Then check the major progressed aspects of the planet ruling the house to see if the astrological activity occurred at the time of the event. When proper relations are established, it is clear that correct combinations apear.

5. *Prenatal Epoch.* At this point after establishing a general picture, it may be possible to apply the prenatal epoch, a popular method of rectification to find the exact time of birth. However, our research activities show quite conclusively that it is valueless unless the time of birth is ascertained within half an hour of the correct time by some other method. In fact, about seven out of every ten birth-charts can be rectified by the prenatal epoch when the time of birth is known within *thirty minutes.*

6. *Precise event check.* Each reenforced major progressed constant of an event or disease is always released by a transit aspect heavier than from the Moon to one of its four terminals at the time the event occurs or the disease develops. And an independent minor progressed aspect is always released by a transit aspect to one of the birth-chart or major progressed

terminals influenced by the minor progressed aspect at the time the event takes place.

The above research finding enables us to eliminate the possibility of a given event or the commencement of a given disease—even during those periods while the major progressed Constants are within one effective degree of orb, and all are being reenforced by minor progressed aspects—except during those limited periods when ALL these major progressed constants are also being released by transit aspects.

This information assists greatly in rectifying birth-charts, for the event will not occur on one of the peaks of power indicated when ALL the major progressed constants are reenforced by minor progressed aspects. But it will occur during one of these peaks of power ONLY at a time when there is also a transit aspect releasing ALL the reenforced major progressed constants of the event or disease.

Therefore, where the exact date is known, the majors, minors and transits should be active relative to the constants for each given event. To find the exact degree on the Midheaven and Ascendant, events should be selected that cannot adequately be explained by the progressed planets. After the chart is rectified as closely as possible, the prenatal epoch may be applied as well.

7. *Time of Day of Trial Chart.* When a series of events check with given constants, the relation of planets, houses and signs should coincide with the known factors in the native's life as recorded in his case history. The trial chart has some degree and minute of a sign on the Ascendant. To find what time during the day of birth the chart represents, work backward through the rules for chart erection—starting with the Ascendant in the Tables of Houses and finally arriving at the Local Mean Time of birth.

Now the chart is ready for future checking of daily, weekly and monthly events. This trial and error method requires persistent work and careful checking, but if carried out thoroughly we find it gives a reliable chart—more reliable than one rectified by any other method.

95. CHIEF RULER OF HOUSE AND HOUSE CUSP ORBS.

Probably due to lack of precision in timing births, consequently obtaining incorrect degrees on various house cusps, together with the observed affect of planets as many as 10 degrees from Midheaven into ninth house and from Ascendant into twelfth house, an opinion gained ground in astrological circles that when a planet was within 5 degrees of a house cusp it not only mapped activity concerning things ruled by the house it was in, but also mapped activity—to some extent—of things ruled by the other house.

Yet the results of an analysis by the research department of natal and progressed positions in many thousands of charts forces the conviction that one house is divided from another by an abrupt partition. Even though a planet maps powerful thought-cell activity regarding things ruled by the house it actually occupies by position, it is not associated (except through ruling the cusp sign of the adjoining house or through aspecting a planet in it) with thought-cell activity mapped by the adjoining house, even when within one degree of its cusp.

It was found that a planet commonly maps the most powerful psychokinetic energy through the house it is actually in; the energy mapped being most powerful when the planet is near the cusp, gradually decreasing as it moves further into the house, and weakest just before crossing over the cusp into the adjoining house.

Cusps representing Midheaven and Ascendant—one acting as a broadcasting antenna, the other a ground-wire—map energies functioning after the manner of those mapped by the planets. Therefore, a planet well over into the ninth house maps an association with Midheaven because it conjuncts it, and sometimes a planet as many as 11 degrees into the twelfth house conjuncts Ascendant. (See Orb Table.)

Extensive experience leads to the conviction that not merely in the birth-chart but also by progression the Midheaven and Ascendant should be considered in the same manner as the planets. It is unnecessary to consider aspects to any other house cusps or to progress the cusps of other houses aside from the Midheaven and Ascendant, both allowed the orb of a planet in an angle.

Birth-chart house cusps map divisions (astral membranes) between one compartment and another in the astral body. While Midheaven and Ascendant are progressed in the Hermetric system, there is no progressed-chart, in the sense of progressing all house cusps. Research revealed that progressed aspects invariably work out in terms of birth-chart houses through which the planets are progressing or which they rule.

When only one planet is located in a certain house, it is chief ruler. The planet ruling the house-cusp sign (because the sign sympathetically transmits some of the energy of the planet ruling it no matter where the planet is located) is a secondary ruler, or co-ruler, of the house. When several planets appear in the house (indicating greater psychokinetic energy available to stimulate events concerning the department of life) the planet nearest the cusp is usually chief ruler, and other planets in the house, co-rulers. Of these co-rulers, those closer to the house-cusp are stronger; those farther, weaker. Yet any of these planets in a house map stronger stimulation of things ruled by the house than does the planet that is co-ruler merely because it rules the house-cusp sign. However, the chief ruler of a house is always the planet in it having the most astrodynes, or if no planet in the house, the one(s) ruling the house-cusp sign.

96. Planet Combust or Cazimi

A planet is considered Combust when it is within 8 degrees and 30 minutes of the Sun, and in a Degree of Combustion when within 17 minutes of the Sun. According to some writers, when a planet is Combust its influence is burnt up and destroyed, and the planet thus being weakened is considered a great affliction.

On the other hand, it is said that a planet Cazimi is greatly strengthened. A planet is Cazimi when within 17 minutes of the Sun's center. As the disc of the Sun occupies approximately half a degree, a planet Cazimi sends its rays to the earth along lines identical with those of the Sun so far as the zodiacal degree occupied is concerned, and the result of such overlapping possibly maps a different influence than when the two bodies are slightly farther apart.

In no case did research find the energy mapped by a planet to be weakened as a result of that planet conjuncting the Sun; on the contrary, a planet conjunction Sun was found to map more psychokinetic energy to attract events than when not aspecting the Sun.

A planet strongly conjunct birth-chart Sun was found to coincide with important factors molding life and character of the native. As Mercury is never over 28 degrees from the Sun, it is combust in about one-fourth of all charts, and in a Degree of Combustion in about two-thirds of them. Venus is never more than 48 degrees from the Sun, so is Combust in at least one-seventh of all charts, and in a Degree of Combustion in about one-third of them.

Due to a great deal of research the customary viewpoint concerning a planet Cazimi or Combust is not held in the Hermetic System of Astrology.

97. Mental Antidote

A chemist wishing to change a compound can add one or more other elements, whose union with one or more other elements in the original compound changes its character entirely. Such is the case in obtaining Nitric Acid by adding Water to Nitrogen Pentoxide. If the axiom "As above, so below" holds, the same manifestation should be possible on the astral plane, where the composition of a thought compound in the astral body can be changed by adding one or more other thought-elements.

Research and statistical studies covering thousands of birth-charts and progressions present at the time events occurred prove that any event of outstanding importance to the native is preceded by an increase in the type of thinking attracting the event.

The normal thought-cell activities for each of the twelve

departments of life differ with each person, and determine the normal prominence and fortune of the life in each department. The amount of money earned, honor attained, health, or success in love normal to one person may be far above that of another; because the first person's thought-cells are capable of releasing desire-energies in sufficient volume, intensity and harmony to attract such fortune. The norm for this person may equal the temporary attainment of the other's most fortunate period in life. Degree of attainment depends upon the thought-cell activity relative to the department of life.

But whatever the normal fortune or misfortune due to habitual thought-cell activity and method of thinking accompanying it, we now know that events considered important by the native (because they are somewhat better or worse than his norm) never enter his life except at times when his thoughts are stimulated in a certain way, as mapped by a major progressed aspect involving the birth-chart house ruling the department of life affected. Such a progressed aspect, as much observation demonstrated, is invariably accompanied by a definite change in the kind or intensity of thinking, unless this change is recognized and an effort is made to redirect the thoughts. By deliberately cultivating another type of thinking than that mapped by the progressed aspect, a different type of event or condition is attracted into the life.

An afflicted birth-chart or progressed planet maps an urge with thought-cell activity tending to attract discord. If the discordant energy is freed, mental poisoning results, attracting misfortune. On the physical plane an antidote is a remedy for counteracting the effects of poisoning, while on the astral plane, a mental antidote serves the same purpose and assists in altering the direction of thought-cell desires—otherwise discordant—into harmonious channels. Thus the mental alchemist employs methods of the material chemist, applying new elements. His degree of success is measured by his ability to associate harmony and intensity with the application of correctly induced emotion and directed thought.

Each planet, embracing one of the ten thought families,

maps a different type of thinking: Sun, power thoughts; Moon, domestic; Mercury, intellectual; Venus, social; Mars, aggressive; Jupiter, religious; Saturn, safety; Uranus, individualistic; Neptune, utopian; Pluto, universal welfare. Each of these thought families may express their energies discordantly or constructively, direction determined by the native's own thinking.

Discordant thought-cells, mapped by an afflicted planet, can be reconditioned when thoughts and experiences of the antidote's nature are substituted for the old type of habitual thinking and feeling.

Energies ruled by the Sun need no mental antidote, and any constructive thoughts of the other planetary families added to Sun thought-cells are beneficial to them.

Constructive thought-elements ruled by Moon or Pluto are the mental antidote for discordant Mars energies, and constructive Mars thought-elements are the mental antidote for Moon or Pluto energies.

Constructive thought-elements ruled by Jupiter are the mental antidote for discordant Mercury or Uranus energies, and constructive Mercury or Uranus thought-elements are the mental antidote for discordant Jupiter energies.

Constructive thought-elements ruled by Venus or Neptune are the mental antidote for discordant Saturn energies, and constructive Saturn thought-elements are the mental antidote for discordant Venus or Neptune energies. In either case Sun thought-elements should also be added to counteract the natural negative nature of all three planets.

To apply mental antidotes, constructive thought-elements should be (1) substituted for the discordant thoughts to be reconditioned; (2) used in a discordant environment belonging to this planetary family; and (3) used whenever the planet mapping the discordant thoughts is involved in a progressed aspect.

Birth-chart Constants showing a predisposition toward specific diseases and difficulties can be ascertained and their corresponding thought-compounds reconditioned. But the person aspiring to live life at its highest is not content merely to thwart some disease, or merely to make a success in business or matrimony. Instead, he desires the utmost from life, to build for himself as complete a success—physically, mentally and spiritually—as possible.

Simply applying a Mental Antidote to some one or two troublesome thought-cell groups is not enough. What is required is a thorough overhauling of the character by properly strengthening all weak parts and adding necessary new parts to attract desired conditions and events. This means a complete rebuilding of character—the only possible action insuring a correspondingly better grade of conditions and events attracted. Astrology and psychoanalysis, using control technique (See Report No. 100), make possible the highest type of living.

98. ENDOCRINE GLAND RESPONSE TO THOUGHT-CELL ACTIVITY

Man is a two-plane organism whose outer-plane body is composed of physical cells and structures and whose inner-plane body is composed of thought-cells and structures. He reacts about equally to his outer-plane environment and to his inner-plane environment. His physical body reacts to heat, cold, rain, wind, food, people, and inanimate objects; and his states of consciousness coincident with these physical responses add energy to and somewhat change the thought-cell organization of his finer form.

A great deal of carefully checked observation indicates that he responds to the inner-plane weather mapped by progressed aspects as much—probably more—as he responds to the outer-plane weather. The thoughts of people and inner-plane entities, and inner-plane radiations from objects add their energies to certain groups of thought-cells within his finer form, giving them greater activity than usual and enabling them through his electromagnetism to influence his physical body.

One of the means through which the chemistry of the physical body is so quickly altered by such thought-cell activity is the system of endocrine glands, ruled by the Sun. These are the glands of internal secretion. They manufacture complex compounds called hormones chiefly from proteins attained from the blood and lymph. Each endocrine gland manufactures its own hormone or hormones, differing in chemistry and function from the hormones manufactured by the other glands. These hormones are not emptied into a duct which carries them to the region to be activated, but instead are liberated directly into the blood stream. That the hormones may thus find easy access to the region affected, the endocrine glands usually have a rich supply of blood vessels.

Endocrine gland activity may be stimulated by thought or emotion, by the accelerated thought-cell activity within the soul mapped by a progressed planetary aspect, or by the hormones of other endocrine glands. Their ready response to conscious thought, or to unconscious thought-cell activity inaugurated by a progressed planetary aspect, is accomplished by nerves carrying electrical impulses to the glands. Yet it has only been since 1936 that nerve currents have been recognized by science to be electrical, and only since 1922 that science knew anything about the endocrine control exerted over the body.

When an endocrine gland is stimulated by a thought, emotion or thought-cell activity, its hormone enters the blood and is carried along in the general circulation to be distributed to all parts of the body. All the cells and tissues are bathed in the fluid carrying it, and some cells, organs and other glands respond to it in a marked way, even though distantly located in the body.

Among other things these glands—each of which is delicately responsive to the stimulated thought-cell activity mapped by one or more specific planet—determine the size, shape and texture of the body; make for intelligence or its lack; give courage or cowardice; imbue with ambition or saturate with laziness; prompt to moral actions or those immoral; and, in general, force the given outlook upon life. And they determine the chemistry of the body at a given time; therefore, just what foods are needed to keep or restore health.

Lest the implication be taken that progressed aspects alone

are instrumental in accelerating specific glandular secretions, it should be remembered that thought-cell activity such as is ruled by any planet can be increased by appropriate conscious thinking. Conscious thinking is boundary-line thinking, imparting some of its energy to corresponding thought-cells within the soul, and some more directly, by means of nerve currents, to the glands. Also, heat or cold, and the foods taken into the system act from the outer plane to increase or depress glandular activities. Regardless of the cause activating the glands, hormones they secrete chiefly determine the body chemistry and what food it needs at any given time.

Because each endocrine gland is so responsive to the thought-cell activity of specific planetary types, its tendency to be influenced from normal at predetermined times is mapped by progressed aspects indicating the type of planetary energies to which it responds. Because the prominence and aspects of the birth-chart planets map relative amounts of thought-cell activity of various types at birth, the general predisposition of the individual to over or under activity of certain glands may be judged from the birth-chart.

Not only do some endocrine glands secrete more than one hormone, but several of them respond with almost equal alacrity to types of thought-cell activity ruled by more than one planet.

All the hormones of these glands have not been isolated as yet, and the front pituitary, which a few years ago was considered to secrete only one hormone, is now known to manufacture not less than twelve; even as vitamin B also has been found to comprise many vitamins, twelve of which have now been discovered, the whole of which is termed vitamin B complex.

Long and close observation by The Church of Light Astrological Research Department has established the relationship between hormone activity and definite types of thinking mapped by prominent planets either in a birth-chart or by progression as follows:

SUN thought-cells influence two of the front pituitary gland

—one stimulating the thyroid gland and otherwise contributing to energy transformation necessary for sustained effort, and the other stimulating the adrenal cortex and with that gland determining the presence of courage—and the secretion of thyroxin, the thyroid hormone of energy production. These three hormones respond either to thought-cell activity stimulated by progressed aspects involving the Sun, or to conscious thinking of the POWER type.

MOON thought-cells influence the two back pituitary hormones, the thymus hormone and the hormones of the alimentary tract. One back pituitary hormone regulates woman's periodic functions. The other, pituitrin, causes water retention in the body and contraction of all arteries except those in the kidneys. It also causes contraction of the plain muscles of bowels, bladder, womb and other organs; is sometimes used in childbirth to contract the womb; has something to do with the metabolism of the carbohydrates; and its deficiency leads to obesity. The thymus hormone tends to delay maturity. These hormones respond either to thought-cell activity stimulated by progressed aspects involving the Moon, or to conscious thinking of the DOMESTIC type.

MERCURY thought-cells influence the hormone of the parathyroid glands and one of the hormones of the front pituitary. Parathyrin, the hormone of the parathyroid glands, controls the body's calcium metabolism, and in combination with other hormones ruled by Mercury, Uranus, Neptune and Pluto, controls the type of electrification. The front pituitary hormone responding to Mercury thought-cells influences cerebral activity. These two hormones respond to thought-cell activity stimulated by progressed aspects involving Mercury, or to conscious thinking of the INTELLECTUAL type.

VENUS thought-cells influence the hormone of the thyroid gland and those of the gonad glands. Testosterone is the male sex hormone of the testes. Alphaestradoil has been found to be the active principle of estrone, the female hormone of the ovaries, now commonly referred to as the estrogenic hormone. These gonad hormones contribute to virility and rejuvenation.

The corpus leuteum hormone of pregnancy, progestin, also responds to Venus thought-cells. The thyroid secretion, thyroxin, is the hormone of energy production. These hormones respond either to thought-cell activity stimulated by progressed aspects involving Venus, or to conscious thinking of the SOCIAL type.

MARS thought-cells infiuence hormones of the adrenal glands and the gonad glands, Testosterone, the male gonad gland hormone, has been given great publicity as the generator of male virility and a rejuvenator of the whole body. The adrenal glands secrete two hormones: (1) Adrenalin, secreted by the adrenal medulla, hormone of emergency energy, and (2) cortin, secreted by the adrenal cortex, hormone of courage. Both adrenal hormones neutralize toxicity of the blood and are the chief chemicals used by the body to fight infection. These sex hormones and adrenal hormones respond either to thought-cell activity stimulated by progressed aspects involving Mars, or to conscious thinking of the AGGRESSIVE type.

JUPITER thought-cells influence the hormone of the pancreas, insulin. Sugar and starch can be used as fuel by the body, or as the kindling necessary to burn fat, only when insulin is present. The liver, central banking system of the body, stores reserve fuel in the form of animal starch, or glycogen—a process of storage depending upon the insulin supply. This hormone responds either to thought-cell activity stimulated by progressed aspects involving Jupiter, or to conscious thinking of the RELIGIOUS type.

SATURN thought-cells influence one front pituitary hormone which governs the growth of the skeleton and supporting tissues. And equally with Mars, Saturn thought-cells influence the secretion of adrenalin, the hormone of emergency energy. These two hormones respond either to thought-cell activity involving Saturn, or to conscious thinking of the SAFETY type.

URANUS thought-cells influence one front pituitary hormone related to original thinking and influence the secretion of the parathyroid glands. Parathyrin, the hormone of these glands, controls the calcium metabolism of the body, and in combination with other hormones ruled by Mercury, Uranus, Neptune and Pluto, controls the type of electrification. These hormones respond either to thought-cell activity stimulated by progressed aspects involving Uranus, or to conscious thinking of the INDIVIDUALISTIC type.

NEPTUNE thought-cells influence the hormone of the pineal gland and the hormone of the parathyroid glands. The pineal hormone prevents precocious maturity, and it neutralizes and depresses the secretion of adrenalin and cortin. Parathyrin, the parathyroid hormone, controls the calcium metabolism of the body, and in combination with other hormones ruled by Mercury, Uranus, Neptune and Pluto, controls the type of electrification. These hormones respond either to thought-cell activity stimulated by progressed aspects involving Neptune, or to conscious thinking of the UTOPIAN type.

PLUTO thought-cells influence the hormone of the pineal gland; one front pituitary hormone relating to cooperative effort; the cortin hormone of the adrenal cortex; and the secretion of the parathyroid glands. The hormone of the pineal gland prevents precocious maturity, and it neutralizes and depresses the secretion of adrenalin and cortin. Cortin is the hormone of courage, neutralizes toxcity of the blood, and is one of the chief chemicals used by the body to fight infection. Parathyrin, the parathyroid hormone, controls the calcium metabolism of the body, and in combination with other hormones ruled by Mercury, Uranus, Neptune and Pluto, controls the type of electrification. These hormones respond either to thought-cell activity stimulated by progressed aspects involving Pluto, or to conscious thinking of the UNIVERSAL WELFARE type.

The cells of the body are miniature batteries generating electricity and radiating high-frequency energy of lightning which fixes the nitrogen that plant life we consume takes from the soil. This is in accordance with Einstein's Law of Equivalence; the energy of an atom is given out in the same quanta as those received by the atom. And of the cells of the human body, those of the nervous system are best adapted to the production of short-wave radiations, as well as to carrying electricity. But thought-cell activity determines, through the action of endocrine gland hormones, the frequency of the wave lengths radiated at a given time. And this thought-cell activity, in turn, may arise chiefly from the stimulation of factors within the soul by the energies mapped by a progressed aspect, or by a consciously selected mood and trend of thought.

99. ASTROLOGY, HEALTH AND DIET

Each individual has certain functions to perform on earth, and the machine tools he must employ in doing whatever his work may be are his mind and his body. He can do his work to best advantage only if his mind and body are kept in first rate condition.

Residing on the inner plane is his mind, consisting of all his past experiences, including those before he was born into human form. It not only embraces whatever knowledge he may possess—and upon this knowledge depends his ability—but also the way past experiences have been organized as harmonious or discordant thought-cells, and the dominant vibratory rate which determines the inner-plane level on which his mind functions and thus his spirituality.

The physical body, including the brain, resides on the outer plane and embraces various types of physical tissue and various organs—the way they are related to each other determining their degree of coordination. While some things are accomplished by the psychokinetic power of the mind, the accomplishment of each person, in large measure, depends upon thoughts reaching objective consciousness through the brain and determining what the body does—its behavior.

The mind, which is one machine tool residing on the high-velocity inner plane, is attached to and functions by psychokinesis through the other machine tool, the body. This ability to function through such a complex form by means of the non-physical power of the mind has required long training of evolution, during which the mind step by step, through being attached successively to higher life-forms, has developed the necessary knowledge and psychokinetic power.

Neither of the machine tools, on which each individual must depend for both accomplishment and the events which enter his life, is perfect at birth. He has to learn physical coordination in order to walk, talk, and do more complex physical things. The physical tool must be given proper foods and a congenial environment if it is to be kept healthy and able to do effective work.

To direct his physical machine properly, he must gain knowledge and store it for future use in his inner-plane tool, the mind. This inner-plane tool, to be properly effective, must also receive training and proper foods. The discords in it—thought-cells using their psychokinetic power to attract misfortune—must be trained to have different desires. To that end they must be reconditioned. As the nutriments of mind are mental states, and as health of the body and the events coming into life are profoundly influenced by the health of the mind, it is fully as important to feed the mind properly as it is to properly feed the physical body.

Certain outer-plane environmental factors such as heat, cold and weather, affect the physical health. And certain factors of the inner-plane environment, such as the inner-plane weather mapped by astrology, affect the mental health. One needs different physical food when the outer-plane temperature is below zero than when it is 100° in the shade. And different mental food is required when cold Saturn dominates the inner-plane weather than when Mars raises the temperature so high anger tends to flame on the slightest provocation.

As the condition of the outer-plane machine markedly affects the condition of the inner-plane machine, and the inner-plane machine condition strongly affects the outer-plane machine condition, it is important that each be given the particular foods whose need is indicated by the birth-chart. Each should be given the foods needed during specific periods indicated by inner and outer-plane weather to put these two machines in as good running condition as possible.

Because the individual is not born until the inner-plane

weather corresponds in general to his inner-plane, or thought-cell organization, his birth-chart indicates along broad lines the types of mental food needed to put his mental machine in order and the types of physical food that will assist the mental machine to put his physical machine in good running order.

While he must learn through experience which physical foods agree or disagree with him, research pointed out that certain physical foods and certain mental foods are especially and urgently needed if in his chart a given planet is heavily afflicted. The chart of birth shows the PREDISPOSITION toward diseases of a particular type. The most powerful and discordant birth-chart planet shows the type of disease toward which he has the strongest predisposition. The next most powerful and discordant planet shows the type of disease toward which his predisposition is next strong, and so no.

The birth-chart constants of different diseases are given, as statistically ascertained from the analysis of the birth-charts of those who have experienced the disease, in previous research reports. It is well to become familiar with these constants so that the disease toward which one has a predisposition can be known, and precautionary actions taken to prevent their development. These precautionary actions in particular embrace both the physical and mental diet needed when planetary constants are afflicted.

A person should also know what progressed aspects coincide with the development of the disease toward which he has a predisposition. Before these progressed constants are within one degree of perfect, and during the time they are within the one effective degree orb of perfect, he should take special pains to follow both the mental and physical diet advocated for afflictions of each planet involved.

Furthermore, even though a planet heavily afflicted by progressed aspect is not a constant for the disease toward which he has a predisposition, while the progressed affliction is within the one degree of effective orb, he should use both the physical and mental diet indicated. In that way he will not only do much to prevent the stimulation mapped by the progressed affliction from decreasing the efficiency of his two machine tools, but he will mitigate the severity of the events that otherwise are likely to come into his life during this period.

To use the mental and physical diets best suited to keep both machines in good running order, he must familiarize himself with his birth-chart, progressions, general principles of a healthy diet, the function of each of the various types of food, vitamins, and the importance of a proper alkali-acid balance.

To recondition the thought-cell organization of the finer form, it is necessary to induce appropriate emotions; for it is the feeling accompanying thoughts that determines whether they build harmonious or discordant thought-cells. And it is the psychokinetic activity of the thought-cells that brings most events of consequence into the life.

Afflicted Planet	**Physical Diet**	**Mental Diet**
Sun	Iodine, Manganese, Vitamins B Complex and A.	Harmonious thoughts of vitality and power.
Moon	Proper Water-salt Balance, Vitamins B-2.	Harmonious Mars thoughts.
Mercury	Calcium, Vitamin B-1, Vitamin D or Sunlight.	Harmonious Jupiter thoughts.
Venus	Iodine, Copper, Vitamins A and E.	Harmonious Sun and Saturn thoughts.
Mars	Iron, Low Protein (but a variety), Vitamins A, B-1 and C.	Harmonious Moon thoughts.
Jupiter	Sulphur, care with sugar and fat.	Harmonious Mercury thoughts.
Saturn	Variety in Mineral Salts, Vitamins, Proteins.	Harmonious Venus and Sun thoughts
Uranus	Calcium, Vitamin B-1, Vitamin D or Sunlight.	Harmonious Jupiter thoughts.
Neptune	Energy food, Calcium, Less Protein, Vitamins D and B-1.	Harmonious Saturn and Sun thoughts.
Pluto	Protein, Iron, Calcium, Vitamins A, C, D, B Complex.	Harmonious Mars thoughts.

100. HOW TO RULE THE STARS

Each birth-chart planet broadcasts energy of a special type, but planets are neither responsible for character with which a person is born nor for events which later enter his life. Nor are progressed planets responsible for the events of a person's life. True, they determine, in large measure, his astral environment at any given time, but how that invisible environment affects him is determined by his reaction to it—not by the planets.

Two things are involved in every condition and event of life: (1) organization of the thought-cells within the astral form constituting character, and (2) environment to which that character reacts.

If the thought organization is changed, even though environment remains the same that which happens will be different. If the thought structure of the astral body is changed, energies mapped by the progressed aspects otherwise indicating misfortune can be diverted into channels that will attract good fortune.

There is a direct relationship between energy and work—a relationship which holds on the four-dimensional plane as well. Events of only so much importance can be attracted by thought organizations with limited energy supply working from the four-dimensional plane, no matter how effectively that energy is directed.

The birth-chart indicates rather closely the amount of thought-cell energy at birth in each compartment of the astral body. However, that desire energy was built into the astral body gradually through states of consciousness associated with things related to various departments of life. Further, the birth-chart shows only that desire energy as it existed at birth. THE FOUR-DIMENSIONAL FORM IS THOUGHT-BUILT AND NOURISHED BY THOUGHTS; THEREFORE, THERE IS NO REASON TO SUPPOSE THAT THOUGHT - BUILDING PROCESSES STOPPED AT BIRTH.

In other words, if sufficient thought and effort is devoted to building new thought-cells and re-organizing old thought-cells in the astral body compartment relating to some department of life, enough energy can be added so that the four-dimensional form will work to attract important events.

The general rule, supported by comprehensive statistical studies, is that the more energy possessed by the thought-organization associated with any given phase of life, the more important—either favorable or unfavorable—the events relating to it become. Those who—consciously or unconsciously—worked energetically to build experiences into their finer bodies relating to a specific thing demonstrated it. They not only worked with things associated with the given department, but repeatedly put a great deal of thought and feeling into the project. All that the birth-chart now indicates has at some time been built into the astral body in the same manner, AND THE PROCESS OF THOUGHT-BUILDING IS AS OPEN FOR USE NOW AS IT EVER WAS.

Energy can be supplied by properly directed intense thought and feeling, but commonly it is furnished by mental stimulation mapped by the planets. Life moves along normally with the thought-cells striving to satisfy conditioned desires. But when a progressed aspect forms within one degree of perfect, resultant activity furnishes more four-dimensional power. And statistical work with events occurring in thousands of people's lives—from their timed birth-charts and progressions—shows that unsuual and important events occur only at times of unusual thought-cell activity.

It is easy to determine when events related to a definite department of life will occur, because a progressed aspect within one degree of perfect maps them. But it is difficult to tell how much benefit or distress will accompany the event under any aspect. Research amply proves that CONDITIONING of thought-cells and thought-structures in the astral body more surely determines the nature of their four-dimensional activity than does the discordant or harmonious energy they receive at any particular time through a progressed aspect.

Five Methods of Control

In ruling the stars, five methods of control fall into three general classifications: (A) Most obvious is through manipulating physical environment and directing physical actions. (B) changing the thought-compound of the stellar cells and structures reached by the planetary energy, in order that any energy reaching them will be utilized in a different type of four-dimensional activity. (Conversion and Metal Antidotes.) (C) Changing the volume of harmony or discord of the energy, to incite more or less activity, or to make the thought-cells feel more disagreeable or more congenial. (Character-Vibrations and Rallying Forces.)

1. *Manipulating the physical environment.* If environmental resistance to a particular event is sufficiently high, thought-cells in the four-dimensional form do not acquire enough energy and the event that they otherwise would attract does not happen. (See B. of L. Course X-I, "Delineating The Horoscope".)

2. *Conversion* is reconditioning of thought-cells—rearranging elements already present to change their feeling—so they can work energetically to attract only favorable conditions. (See B. of L. Course V, "Esoteric Psychology" and XIV, "Occultism Applied".)

3. *Mental Antidotes* are used for adding new thought-elements to convert existing four-dimensional structures into new and beneficial compounds, giving them a different type of activity. Applied effectively, mental antidotes are the most desirable method of control, because their application produces the quickest change in mental function.

4. *Rallying Forces* represent energies mapped by planets in effective orb of aspect at any given time. They may be indicated by the natal chart, by progressions, or both. Since energies of the mind are directed by attention and conditioned by pleasure and pain, Rallying Forces may be used in any desired channel by directed thinking and induced emotion. Energy expended in constructive effort cannot be used destructively at the same time. Through this method of control energies mapped as discordant may be channeled into harmonious expression. (See B. of L. Course IX, "Mental Alchemy".)

5. *Character Vibrations* are not those of personal thoughts nor those corresponding to the planets, but include thoughts of our human associates as well as astral vibrations radiated by all objects and conditions of our environment. Names, numbers, tones, colors, types of environment, etc., radiate definite vibratory rates. They may be selected for the specific effect they will have in furnishing a definite kind of energy to the thought-cells. (See Course VI, "The Sacred Tarot" and XVIII, "Imponderable Forces".)

Research provides us with the following summation: What comes into our life is not due to the positions of the stars either in the birth-chart or after birth; it is chiefly due to four-dimensional activity of thought-cells within our astral bodies. If we would rule our stars, therefore, in addition to manipulating the physical environment intelligently, we must change thought-cell activity. Then the change in our characters can not fail to bring a corresponding change in events and conditions attracted. By employing methods of control, desires of thought-cells can be reconditioned, and effective control of energies indicated, as well as their usage in accomplishment of constructive work, will equal the highest type of success.

Character Determines Destiny. "While there is a certain range of vibratory rates within which the similarity between those of the character and those in the sky is close enough that the child then born will live, nevertheless this similarity—as careful analysis of tens of thousands of birth-charts by our Astrological Research Department proves—is always close enough that the outstanding factors of character are accurately mapped in the birth-chart.

"And as not only the abilities, but every event in life, is an expression of, or is attained by, those thought-organizations which comprise the character, the birth-chart gives a clear picture of the life if nothing special is done to change the character. That is, as all that happens is the result of character, the only manner in which the destiny can be changed is to change the character. Furthermore, as destiny is the outcome of character, and through intelligently directed effort the character can be changed, the life indicated by the birth-chart—which is merely a map of the character with which the individual is born—can be markedly altered in any direction desired."

—From "The Laws of Occultism"
By C. C. Zain

IV

PRELIMINARY ASTROLOGICAL RESEARCH

WE DO NOT CONSIDER the limited number of charts employed in the study of the following items sufficient to afford positive constants. The determination of such constants awaits more data and further research.

A discussion of the 141 diseases that follow, as well as those diseases that appear in the Statistical Research Reports in this book, will be found in "STELLAR HEALING", by C. C. Zain (Brotherhood of Light Course XVI).

The information about each condition is presented in a uniform pattern: the name of the disease, its definition from Gould's Medical Dictionary, birth-chart clues called natal, and the relation of progressions to birth-chart called Progressed clues.

1. *ABDOMINAL TROUBLES*

Definition: pertaining to the cavity in the body between the thorax and the pelvis.

NATAL: in nine cases out of ten an afflicted planet actually in the zone where the disease manifests, and in the other one case out of ten the planet ruling the zone, but elsewhere located, severely afflicted. Cancer rules the stomach, Virgo rules the duodenum and intestines, Libra rules the upper kidneys, the ovaries and internal generative organs, and Scorpio rules the bladder, sigmoid flexure, prostate gland, uterus and external generative organs.

PROGRESSED: depends upon the nature of the disease but a planet in the zone receiving an adverse aspect is the most common factor. When an operation is performed, there is always a progressed aspect involving Mars. The organs and regions ruled by Libra and Scorpio are so closely associated in their functions that an affliction from one commonly affects the other.

2. *ABSCESS*

Definition: a circumscribed cavity containing pus.

NATAL: prominent and usually afflicted Neptune or Pluto, and a prominent and usually afflicted Mars. Afflictions to either Neptune or Pluto depress the production of cortin, the strongest chemical with which the body fights toxins and infection. Mars afflictions tend to exhaust both adrenalin and cortin.

PROGRESSED: usually afflictions, involving both Mars and Neptune, or involving both Mars and Pluto. The low output of cortin hormone results in the incomplete metabolism of protein foods, with an accumulation of toxins. Mars coincides with inflammation and the attraction of infection.

3. *ACIDOSIS*

Definition: acid intoxication caused by an abnormal production and faulty elimination of acids in the body.

NATAL: either a prominent Jupiter or a prominent Saturn or both, usually afflicted.

PROGRESSED: almost any progressed aspect affecting the health adversely through surfeit, faulty elimination, food deficiency, or worry, anger or constant irritation depleting the adrenalin supply, increases this condition. But particularly any aspect to the planet mapping, by its prominence and birth-chart aspects, the predisposition to the difficulty.

4. *ACNE*

Definition: inflammation of the sebaceous glands from retained secretion.

NATAL: Venus discordant; usually heavily afflicted.

PROGRESSED: aspect involving Venus or Mars, usually both.

5. *ADENOIDS*

Definition: enlargement of pharyngeal tonsil (nasal).

NATAL: affliction in Libra or Scorpio, or their rulers severely discordant, especially if at the same time there is an affliction in Aries.

PROGRESSED: aspect involving Venus or Mars, usually an affliction.

6. *ADHESIONS*

Definition: healing by granulation of two surfaces or parts.

NATAL: prominent and usually discordant Saturn, especially when Saturn is located in the zodiacal sign mapping the afflicted part of the body.

PROGRESSED: a powerful progressed aspect involving Saturn, usually an affliction.

7. *ALCOHOLISM*

Definition: symptoms of the excessive and/or habitual use of alcohol.

NATAL: prominent and usually heavily afflicted Mars. An aspect involving Mars and the Moon, especially an affliction, increases the predisposition. Generally those who became habitual drunkards also have a prominent and usually discordant Neptune.

PROGRESSED: as aspect, usually an affliction, involving Mars; and commonly an aspect involving Neptune also.

8. *AMNESIA*

Definition: a loss of memory.

NATAL: prominent and usually afflicted Mercury and Moon.

PROGRESSED: afflictions involving Mercury or Moon and heavy afflictions involving Neptune or Pluto at the same time.

9. *ANEMIA*

Definition: a deficiency of blood or of red corpuscles.

NATAL: Neptune and Saturn prominent, and usually severely afflicted. Mars also afflicted.

PROGRESSED: an aspect involving Mars, and usually one involving Saturn or Neptune, at the same time there is a heavy progressed affliction. Jupiter is usually, but not always, involved.

10. *APOPLEXY*

Definition: paralysis from rupture of a cerebral vessel.

NATAL: afflictions in Libra or Scorpio. Venus and Mars may also be afflicted, and Sun prominent.

PROGRESSED: affliction involving Mars.

11. *ARTERIOSCLEROSIS*

Definition: hardening of the arterial walls.

NATAL: Jupiter and the upper-octave planets prominent and usually afflicted.

PROGRESSED: an aspect, usually an affliction, involving one or more of the three upper-octave planets, Uranus, Pluto or Neptune.

12. *ARTHRITIS*

Definition: inflammation of a joint.

NATAL: Saturn and Uranus prominent, usually severely afflicted.

PROGRESSED: an aspect, usually an affliction, involving Saturn.

13. *ASTHMA*

Definition: paroxysmal, difficult or labored breathing, with oppression.

NATAL: Mercury afflicted, and one or more of the upper-octave planets prominent and afflicted. Either afflictions in Gemini, or Mercury receiving heavy aspects.

PROGRESSED: an aspect involving Mercury or the Gemini plant.

14. *ATROPHY*

Definition: wasting of a part from lack of nutrition.

NATAL: Saturn prominent and afflicted.

PROGRESSED: heavy afflictions to Saturn, or less severe aspects to Saturn with heavy Rallying Forces present.

15. *AUTOINTOXICATION*

Definition: a morbid condition produced by poisonous products elaborated within the body.

NATAL: Saturn and Mars prominent and usually afflicted, and either Neptune or Pluto also prominent and usually afflicted.

PROGRESSED: an aspect, usually an affliction, involving Mars.

16. *BED WETTING (Uroclepsia)*

Definition: unconscious discharge of urine.

NATAL: heavy affliction in Libra, or less commonly in Scorpio; or a prominent and afflicted Uranus.

PROGRESSED: aspect involving planet in Libra or Scorpio, or affliction involving Uranus.

17. *BILIOUSNESS*

Definition: condition marked by constipation, headache and loss of appetite due to excess of bile.

NATAL: Jupiter prominent and afflicted, or heavy afflictions involving the Moon or an afflicted planet in Cancer.

PROGRESSED: aspect, usually an affliction, involving Jupiter, Saturn or the Moon.

18. *BLADDER TROUBLE*

Definition: pertaining to the sac-like receptacle for the urine.

NATAL: an affliction in Scorpio, or less frequently a severe affliction involving Mars or Pluto.

PROGRESSED: an aspect to the afflicted planet in Scorpio, or a heavy affliction involving Mars or Pluto.

19. *BLEEDING*

Definition: a letting of blood.

NATAL: a prominent and afflicted Saturn and Neptune.

PROGRESSED: an aspect involving Mars, and commonly an aspect involving Saturn or Neptune also.

20. *BLINDNESS*

Definition: an absence of vision.

NATAL: severe affliction involving Moon and Mars. If Mars is prominent and afflicted, a severely afflicted Moon may cause trouble with the eyes; even a powerful harmonious aspect between Moon and Mars may map less severe difficulties if the Moon is otherwise severely afflicted. In cataract, Neptune is also involved.

PROGRESSED: a Mars aspect with severe Rallying Forces.

21. *BLOOD POISONING*

Definition: absorbtion of toxins into the blood.

NATAL: Mars, and Neptune or Pluto heavily afflicted.

PROGRESSED: an aspect involving Mars and an aspect involving Neptune or Pluto, at the same time there are heavy Rallying Force afflictions.

22. *BOWEL TROUBLE*

Definition: pertaining to the intestine.

NATAL: an afflicted planet in Virgo, or much less commonly a severely afflicted Mercury.

PROGRESSED: an aspect involving the Virgo planet, or involving Mercury at the same time discordant Rallying Forces are present.

23. *BRAIN FEVER*

Definition: inflammation of the large mass of nerve-tissue contained in the cranium, especially the cerebrum.

NATAL: usually an affliction in Aries, but at times only an affliction to Mercury; and a prominent and afflicted Mars.

PROGRESSED: some aspect involving Mercury or the Aries planet, usually both; and at the same time an affliction involving Mars.

24. *BRAIN, WATER ON (Hydrocephalus)*

Definition: a collection of water in the head. Dropsy of the brain.

NATAL: usually an affliction in Aries, but at times only an affliction to Mercury; and a prominent Moon afflicted, involving negative planets.

PROGRESSED: some aspects involving Mercury or the Aries planet, and usually an aspect involving a negative planet.

25. *BRIGHT'S DISEASE*

Definition: an acute and chronic diffuse disease of the kidneys, usually associated with dropsy and albumin in the urine.

NATAL: usually an afflicted planet in Libra or Scorpio, and Venus or Mars, usually both, severely afflicted.

PROGRESSED: an affliction involving Venus or Mars, more commonly Venus, at the same time there are strong Rallying Force afflictions.

26. *BRONCHITIS*

Definition: inflammation of the bronchial tubes.

NATAL: Gemini highly sensitive due to an afflicted Mercury, or more commonly Uranus, Neptune or Pluto in Gemini.

PROGRESSED: an aspect involving Mars, particularly if the Gemini planet or Mercury is at the same time afflicted.

27. *BUNION*

Definition: a swelling of a bursa (small sac interposed between movable parts) of the foot.

NATAL: affliction in Pisces, or Neptune prominent and usually afflicted.

PROGRESSED: an aspect involving the afflicted Pisces plant or Neptune.

28. *BURNS*

Definition: lesion of tissue from dry heat or flame.

NATAL: heavily afflicted Mars and commonly Saturn and Uranus prominent also.

PROGRESSED: an aspect, usually an affliction, involving Mars.

29. *CARBUNCLE*

Definition: a large circumscribed inflammation of the subcutaneous tissue.

NATAL: Jupiter afflicted and Mars prominent.

PROGRESSED: an aspect involving Mars, usually an affliction.

30. *CATARACT*

Definition: opacity of the crystalline lens.

NATAL: affliction involving Moon and Mars, or Mars prominent and Moon severely afflicted; also an aspect, usually an affliction, involving Neptune and Moon.

PROGRESSED: an aspect involving Mars and an aspect involving Neptune at the same time there are severe Rallying Forces.

31. *CATARRH*

Definition: inflammation of a mucous membrane.

NATAL: one or more upper-octave planet prominent, and Jupiter afflicted.

PROGRESSED: an aspect, usually an affliction, involving the prominent upper-octave planet, or an affliction involving Jupiter.

32. *CEREBRAL HEMORRHAGE*

Definition: a flow of blood from the vessels of the brain.

NATAL: Uranus prominent, usually an afflicted Mars, and commonly an afflicted planet in Aries.

PROGRESSED: an aspect involving Uranus and an aspect, usually an affliction, involving Mars.

33. *CHICKEN-POX (Varicella)*

Definition: infectious, eruptive disease of childhood.

NATAL: Mars somewhat afflicted.

PROGRESSED: an aspect involving Mars.

34. *CHILDBIRTH TROUBLES*

Definition: pertaining to labor and confinement of birth.

NATAL: a malefic planet, or an afflicted planet, in the fifth house, or the ruler of the fifth severely afflicted.

PROGRESSED: an affliction to the ruler of the fifth house.

35. *CHOLERA ASIATIC*

Definition: a malignant form of disease characterized by emesis (vomiting), diarrhea, cramps and prostration.

NATAL: Mars prominent, and Mercury or a planet in Virgo afflicted.

PROGRESSED: an aspect involving Mercury or the Virgo planet and a Mars aspect at the same time.

36. *COLD Common*

Definition: catarrh of the respiratory tract; coryza (inflammation of the nose).

NATAL: negative planets prominent, or a negative planet severely afflicted.

PROGRESSED: an affliction involving either Mars or Saturn; but almost any progressed aspect, if the birth-chart predisposition is pronounced.

37. *COLIC*

Definition: spasmodic pain in the abdomen.

NATAL: a planet afflicted in Virgo; less commonly, a severely afflicted Mercury.

PROGRESSED: an aspect involving the Virgo planet or the severely afflicted Mercury.

38. *COLITIS*

Definition: inflammation of the colon

NATAL: affliction in Scorpio, or less commonly an affliction to Mars.

PROGRESSED: an aspect involving the afflicted Scorpio planet or an aspect involving Mars.

39. *COLOR BLINDNESS*

Definition: Abnormalism or deficiency of color-perception.

NATAL: Moon prominent, afflicted and involved in an aspect with Neptune.

PROGRESSED: this defect is usually hereditary and present from birth.

40. *CONSTIPATION*

Definition: a sluggish action of the bowels.

NATAL: Saturn prominent and involved in any aspect with Mercury, or Saturn in Virgo or in the sixth house. To a lesser extent a prominent Saturn involved in heavy aspects, especially afflictions.

PROGRESSED: any aspect, especially an affliction, involving Saturn.

41. *CONVULSIONS*

Definition: a violent involuntary contraction; a spasm or fit.

NATAL: Mercury prominent and afflicted; also Mars and one or more upper-octave planet (Uranus, Neptune or Pluto) prominent.

PROGRESSED: an aspect involving Mars or Mercury, usually both, and afflictions from Rallying Forces.

42. *CORONARY THROMBOSIS*

Definition: formation of a blood-clot in the heart or in a blood vessel.

NATAL: heavy affliction to a planet in Leo or to the Sun; Venus or Jupiter afflicted, and Pluto or Neptune prominent and afflicted.

PROGRESSED: an affliction involving the Leo planet or Sun, a progressed affliction involving Venus or Jupiter at the same time a progressed aspect involves Pluto or Neptune, and a progressed aspect involving Mars.

43. *CYST*

Definition: a membranous sac containing fluid.

NATAL: Jupiter afflicted, usually in Scorpio or another water sign.

PROGRESSED: an affliction involving Jupiter at the same time there are severe Rallying Forces.

44. *DEAFNESS*

Definition: inability to hear.

NATAL: severe affliction involving Moon and Saturn; Saturn prominent and afflicted and a severely afflicted Moon; or a powerful harmonious aspect involving Moon and Saturn if the Moon is otherwise heavily afflicted.

PROGRESSED: an aspect involving Saturn, with severe Rallying Forces.

45. *DEMENTIA PRAECOX*

Definition: profound mental incapacity occurring in young individuals; includes delusions, progression to imbecility, and is incident to the age of puberty.

NATAL: negative planets prominent and Mercury or Moon, often both, severely afflicted.

PROGRESSED: a powerful aspect involving Mercury or Moon at the same time there are severe afflicting Rallying Forces.

46. *DENGUE FEVER*

Definition: a zymotic (pertaining to changes caused by an organized ferment) disease of tropical and sub-tropical countries, with fever, pain in the bones and an eruption like that of measles.

NATAL: Mars and Venus prominent and usually afflicted.

PROGRESSED: an aspect involving Mars, and an aspect involving Venus, usually afflictions.

47. *DERMATITIS*

Definition: inflammation of the skin; cytitis.

NATAL: Venus afflicted, especially by an upper-octave planet, and one or more upper-octave planet prominent.

PROGRESSED: an aspect to Venus and an aspect to an upper-octave planet.

48. *DIABETES MELLITUS*

Definition: a disease characterized by an excessive flow of sugar-containing urine.

NATAL: Saturn and Jupiter prominent and afflicted, frequently involved in an affliction with each other.

PROGRESSED: an aspect involving Jupiter or Saturn, often an aspect involving each.

49. *DROPSY*

Definition: an effusion of fluid into the tissues or cavities of the body.

NATAL: a prominent and afflicted Moon.

PROGRESSED: an aspect involving the Moon at the same time there are afflicting Rallying Forces, especially an affliction involving the Sun.

50. *DUODENUM TROUBLE*

Definition: pertaining to the first part of the small intestine.

NATAL: affliction in Virgo; less commonly a severely afflicted Mercury.

PROGRESSED: an aspect, usually an affliction, involving Mercury or the planet in Virgo.

51. *DYSPEPSIA*

Definition: impaired or imperfect digestion:

NATAL: affliction in Cancer or Virgo, or Moon or Mercury afflicted.

PROGRESSED: affliction in Virgo or Cancer, or involving Moon or Mercury.

52. *ECZEMA*

Definition: inflammation of the skin with exudation of lymph.

NATAL: upper-octave planets prominent, and Venus prominent and severely afflicted.

PROGRESSED: an aspect involving an upper-octave planet, and an aspect involving Venus, especially an aspect between Venus and an upper-octave planet.

53. *EMPYEMA*

Definition: pus in the pleural cavity.

NATAL: Pluto or Neptune afflicted, and either Mercury severely afflicted or an affliction in Gemini.

PROGRESSED: an aspect involving Pluto or Neptune at the same time there are severe Rallying Forces.

54. *EPILEPSY*

Definition: a nervous disease with loss of consciousness, and tonic and clonic convulsions.

NATAL: affliction involving Mars and affliction involving Venus.

PROGRESSED: an aspect to Mars and an aspect to Venus.

55. *EXOPTHALMIC GOITER*

Definition: an enlargement of the thyroid gland, accompanied by abnormal protrusion of eye-balls and cardiac palpitation.

NATAL: affliction involving Sun and Uranus, or less commonly a prominent Uranus and the Sun severely afflicted.

PROGRESSED: Sun aspecting Uranus, or Sun involved in severe affliction, especially if Venus is involved in an affliction at the same time.

56. *FAINTING*

Definition: a temporary suspension of respiration and circulation.

NATAL: Sun afflicted or a planet in Leo afflicted, and Pluto or Neptune heavily afflicted.

PROGRESSED: an aspect to the Sun or the Leo planet at the same time there is a progressed affliction involving-Neptune or Pluto.

57. *FATTY TUMOR*

Definition: an abnormal swelling or growth of the nature of fat.

NATAL: a prominent and severely afflicted Jupiter.

PROGRESSED: an affliction involving the Sun, and an aspect involving Jupiter.

58. *FIBROUS TUMOR*

Definition: an abnormal swelling or growth of thread-like tissues.

NATAL: Saturn prominent and afflicted. If tumor of the uterus, Venus afflicted and a planet in Scorpio or an afflicted Mars.

PROGRESSED: a discordant aspect involving Saturn.

59. *FOOT TROUBLE*

Definition: pertaining to the organ at the extremity of the leg.

NATAL: afflicted planet in Pisces or a prominent Neptune.

PROGRESSED: a discordant aspect involving the Pisces planet, or a discordant aspect involving Neptune.

60. *FUNGUS GROWTH*

Definition: a form of vegetation deriving sustenance from dead organic matter.

NATAL: prominent and heavily afflicted Neptune.

PROGRESSED: an aspect to Neptune with severe Rallying Forces.

61. *GALL BLADDER TROUBLE*

Definition: pertaining to a pear-shaped sac on the under surface of the right lobe of the liver, the reservior for the bile.

NATAL: a prominent and heavily afflicted Mars.

PROGRESSED: an aspect involving Mars, with severe Rallying Forces.

62. *GALL STONES*

Definition: calcareous concretions in the gall-bladder and its ducts.

NATAL: Mars and Saturn prominent and afflicted, usually aspecting each other.

PROGRESSED: an aspect involving Mars and an aspect involving Saturn, at the same time there are severe Rallying Forces.

63. *GANGRENE*

Definition: the mortification or death of soft tissue.

NATAL: negative planets prominent and Mars or Venus, usually both, afflicted.

PROGRESSED: an aspect involving Mars at the same time there is an affliction involving Pluto, Neptune or Saturn.

64. *GASTRIC ULCER*

Definition: an open sore in the digestive tract.

NATAL: one or more upper-octave planet prominent, Jupiter afflicted and either an afflicted planet in Cancer or a heavily afflicted Moon.

PROGRESSED: an aspect involving an upper-octave plant, an aspect involving Jupiter or a planet in Cancer, and an aspect involving Mars.

65. *GASTRITIS*

Definition: inflammation of the stomach.

NATAL: Moon heavily afflicted, especially if Mars is involved, or a planet in Cancer heavily afflicted.

PROGRESSED: an aspect to Mars at the same time there are heavy Rallying Force afflictions.

66. *GOITER*

Definition: enlargement of the thyroid gland.

NATAL: affliction in Taurus, or Venus and Sun afflicted.

PROGRESSED: an affliction involving the planet in Taurus, or Venus or the Sun.

67. *GONORRHEA*

Definition: a contagious inflammation with a purulent discharge from the genitals.

NATAL: Venus and Mars afflicted, or one of them severely afflicted and the other prominent.

PROGRESSED: a discordant aspect involving Venus or Mars.

68. *HAY FEVER*

Definition: a seasonal disease of the nasal mucous membrane, with coryza, cartarrhal inflammation and lacrimation (an excessive secretion of tears).

NATAL: affliction involving Mercury and an upper-octave planet, or upper-octave planets prominent and Mercury severely afflicted.

PROGRESSED: an aspect involving Mercury, especially an affliction.

69. *HEADACHE*

Definition: a pain in the head.

NATAL: a planet in Aries, especially if afflicted.

PROGRESSED: an aspect involving Mars, and especially involving a planet in Aries.

70. *HEART PALPITATION*

Definition: a fluttering, or abnormally fast beating, of the heart — a hollow, muscular body, center of the circulatory system.

NATAL: Sun afflicted or planet in Leo afflicted, at the same time the constants for indigestion or stomach trouble are present.

PROGRESSED: aspect involving Sun or Leo planet with progressed constants for digestive disturbances.

71. *HEMORRHOID*

Definition: a pile, or small blood-tumor at the anal orifice.

NATAL: both Venus and Mars somewhat afflicted.

PROGRESSED: an aspect involving Venus and an aspect involving Mars.

72. *HERNIA*

Definition: the protrusion of a viscus (organ inclosed within the cranium, thorax, abdominal cavity or pelvis) from its normal position.

NATAL: heavy afflictions from Mars, and usually an afflicted planet in Virgo, Libra or Scorpio.

PROGRESSED: an aspect, usually an affliction, involving Mars.

73. *HIGH BLOOD PRESSURE*

Definition: increase of force exerted by the blood upon the vessel-walls.

NATAL: prominent Sun and heavily afflicted Saturn or a heavily afflicted Mars.

PROGRESSED: an aspect involving the Sun or an aspect involving Mars.

74. *HIVES*

Definition: a vesicular cutaneous eruption with itching and urticaria (nettle-rash).

NATAL: Venus and upper-octave planets somewhat prominent and afflicted.

PROGRESSED: an aspect involving Venus at the same time there is an aspect involving an upper-octave planet.

75 *IMPETIGO*

Definition: an acute pustular inflammation of the skin.

NATAL: upper-octave planet prominent and Mars and Venus afflicted.

PROGRESSED: an aspect involving Mars and an aspect involving Venus at the same time there are severe Rallying Forces.

76. *INDIGESTION*

Definition: impaired or imperfect digestion. (See Dyspepsia)

NATAL: afflicted planet either in Cancer or Virgo, or Moon or Mercury afflicted.

PROGRESSED: discordant aspect involving the Cancer planet or Virgo, or a discordant aspect involving Moon or Mercury.

77. *INFECTION*

Definition: acquiring disease without actual contact with the patient, i.e., by air, water, food, etc.

NATAL: prominent and usually afflicted Mars.

PROGRESSED: an aspect, especially an affliction, involving Mars.

78. *INSANITY*

Definition: mental derangement; madness.

NATAL: severely afflicted Mercury or Moon, usually both.

PROGRESSED: heavy afflictions, and nearly always a progressed aspect involving Mercury or Moon, usually an affliction.

79. *ITCHING*

Definition: an irritable tickling of the skin.

NATAL: Venus prominent and afflicted and an affliction involving an upper-octave planet.

PROGRESSED: an aspect, usually an affliction, involving Venus and an aspect to an upper-octave planet at the same time.

80. *JAUNDICE*

Definition: a yellow coloration of the skin.

NATAL: Mars afflicted and Jupiter or Saturn prominent, usually afflicted.

PROGRESSED: an aspect involving Mars and an aspect, usually an affliction, involving Jupiter or Saturn.

81. *LARYNGITIS*

Definition: inflammation of the upper part of the windpipe, or voice organ.

NATAL: a planet afflicted in Taurus, or Venus afflicted.

PROGRESSED: an aspect involving the Taurus planet or Venus at the same time there is an aspect involving Mars and another discordant aspect.

82. *LEPROSY*

Definition: an endemic, chronic, malignant disease with cutaneous and other lesions, due to bacillus leprae.

NATAL: negative planets, especially Neptune and Saturn, prominent and afflicted.

PROGRESSED: an aspect involving Mars and an aspect involving Neptune or Saturn at the same time severe Rallying Forces are present.

83. *LEUKEMIA*

Definition: a fatal disease of the hematopoietic (blood-making) tissues, with a great increase in the number of leukocytes (white blood-corpuscles) in the blood.

NATAL: Mars afflicted and Saturn and Neptune prominent, usually afflicted.

PROGRESSED: an aspect involving Mars, an aspect involving Saturn, and an aspect involving Neptune, with severe Rallying Forces.

84. *LIVER TROUBLE*

Definition: pertaining to the largest glandular organ of the body, secreting bile.

NATAL: Jupiter prominent and afflicted.

PROGRESSED: an aspect, usually an affliction, involving Jupiter.

85. *LOW BLOOD PRESSURE*

Definition: decrease of force exerted by the blood upon the vessel-walls.

NATAL: Saturn prominent and Sun afflicted.
PROGRESSED: an aspect, usually an affliction, involving Saturn, and an aspect involving the Sun.

86. *LUMBAGO*

Definition: pain in the loins (lower part of the back).
NATAL: a planet afflicted in Libra or Venus severely afflicted; afflictions from Mars and Saturn at the same time.
PROGRESSED: an aspect involving the Libra planet or Venus, and an aspect involving Mars or Saturn, usually both.

87. *MALARIA*

Definition: an infectious disease caused by a protozoan parasite in the blood.
NATAL: Mars prominent and Neptune, Saturn or Pluto prominent and afflicted.
PROGRESSED: an aspect involving Mars and an aspect involving Neptune, Saturn or Pluto, especially Neptune.

88. *MASTOID TROUBLE*

Definition: pertaining to the nipple-shaped cells of the protruding part of the temporal bone.
NATAL: Saturn and Moon afflicted and in aspect with each other, usually a discordant aspect.
PROGRESSED: a discordant aspect involving Saturn or the Moon.

89. *MEASLES (Rubeola)*

Definition: a contagious disease of children, with an erroption of the skin.
NATAL: Mars prominent and Saturn, Neptune or Pluto Prominent.
PROGRESSED: an aspect involving Saturn, Neptune or Pluto.

90. *MENOPAUSE TROUBLE*

Definition: pertaining to end of menstrual or reproductive life.

NATAL: Venus severely afflicted.

PROGRESSED: an aspect, especially an affliction, involving Venus at the same time there are severe Rallying Forces.

91. *MISCARRIAGE*

Definition: the expulsion of the fetus between the fourth and sixth months of pregnancy.

NATAL: a malefic planet in the fifth house, or the ruler of the fifth severely afflicted. Venus or Mars affliction, especially to each other, increases the danger of this mishap.

PROGRESSED: an aspect to the ruler of the fifth house.

92. *MUMPS (Idiopathic Parotiditis)*

Definition: an acute infectious disease marked by swelling of the parotid gland.

NATAL: Mars prominent and Saturn, Neptune or Pluto prominent.

PROGRESSED: an aspect involving Mars and an aspect involving Saturn, Neptune or Pluto, at the same time there are progressed afflictions.

93. *NERVOUSNESS*

Definition: an unsettled condition of the nerves.

NATAL: Mercury or Uranus severely afflicted, especially by an upper-octave planet.

PROGRESSED: an aspect involving Mercury or an upper-octave planet, especially an aspect between Mercury and an upper-octave planet.

94. *NEURALGIA*

Definition: pain in a nerve.

NATAL: Mercury or Uranus severely afflicted.

PROGRESSED: an affliction involving Mercury or Uranus, or an aspect involving Mercury or Uranus with severe Rallying Forces.

95. *NEURASTHENIA*

Definition: exhaustion of nerve force.

NATAL: Mercury or Uranus severely afflicted.

PROGRESSED: an aspect involving Mercury or Uranus at the same time there are severe Rallying Forces, especially involving Saturn, Neptune or Pluto.

96. *NEURITIS*

Definition: inflammation of a nerve.

NATAL: Uranus or Mercury afflicted by Mars.

PROGRESSED: an aspect involving Mars at the same time there is an aspect, usually an affliction, involving Mercury or Uranus.

97. *NEURODERMATITIS*

Definition: a neurotic dermatitis (inflammation of the skin) with itching.

NATAL: prominent upper-octave planet afflicted and Mercury afflicted by an upper-octave planet.

PROGRESSED: Mercury involved in an aspect with an upper-octave planet; an aspect involving Mars; and an affliction involving Venus.

98. *OBESITY*

Definition: fatness; corpulence

NATAL: Jupiter or Moon, often both, prominent and afflicted.

PROGRESSED: an aspect involving Jupiter or to the Moon.

99. *OBSESSION*

Definition: possession by a demon.

NATAL: a prominent and severely afflicted Neptune, Pluto or Moon.

PROGRESSED: an aspect to Pluto at the same time there are severe Rallying Forces, especially involving Neptune or the Moon.

100. *OSTEOMYELITIS*

Definition: inflammation of bone marrow.

NATAL: Mars prominent and Saturn heavily afflicted.

PROGRESSED: an aspect involving Mars and an aspect involving Saturn, at the same time there are severe Rallying Forces.

101. *OVARY TROUBLE*

Definition: pertaining to the organ of generation in the female.

NATAL: Venus afflicted.

PROGRESSED: an affliction involving Venus, or an aspect involving Venus with severe Rallying Forces.

102. *PELLAGRA*

Definition: a disease of doubtful etiology, characterized by digestive disturbances, skin lesions, and nervous symptoms. Now said to be a deficiency disease.

NATAL: afflictions from Saturn.

PROGRESSED: an affliction involving Saturn, Neptune or Pluto, usually Saturn.

103. *PERITONITIS*

Definition: inflammation of the serous membrane lining abdomen.

NATAL: an afflicted planet in Virgo, or Mercury severely afflicted.

PROGRESSED: an aspect, usually an affliction, involving Mars, at the same time there is an aspect involving the Virgo planet or Mercury.

104. *PHLEBITIS*

Definition: inflammation of a vein.

NATAL: Venus prominent and afflicted.

PROGRESSED: an aspect involving Venus and an aspect involving Mars with heavy Rallying Forces.

105. *PITUITARY DEFICIENCY* (*Posterior*)

Definition: a lack of secretion of small reddish gland.

NATAL: prominent and heavily afflicted Moon.

PROGRESSED: progressed affliction to the birth-chart Moon.

106. *PLEURISY*

Definition: inflammation of pleura (serous membrane enveloping the lungs).

NATAL: Mars afflicted and either Mercury heavily afflicted or an affliction in Gemini.

PROGRESSED: an aspect involving Mercury or the Gemini planet and an aspect involving Mars, at the same time there are severe Rallying Forces.

107. *POISONING*

Definition: a venomous or toxic agent in the body.

NATAL: prominent and afflicted Neptune.

PROGRESSED: an aspect involving Neptune with severe Rallying Forces.

108. *PROSTATE TROUBLE*

Definition: pertaining to a glandular body situated around the neck of the bladder in the male.

NATAL: Mars or Pluto afflicted, or an afflicted planet in Scorpio.

PROGRESSED: an aspect involving Mars, Pluto or the Scorpio planet and severe Rallying Forces active.

109. *PSORIASIS*

Definition: a chronic inflammatory skin disease, with scale formation.

NATAL: an upper-octave planet, especially Neptune, prominent and afflicted and Venus afflicted.

PROGRESSED: an aspect involving an upper-octave planet and a discordant aspect involving Venus.

110. *PYLORIC TROUBLE*

Definition: pertaining to the opening of the stomach into the duodenum.

NATAL: an afflicted planet in Virgo, or much less commonly a severely afflicted Mercury.

PROGRESSED: an aspect involving the Virgo planet or Mercury at the same time there are severe Rallying Forces.

111. *RHEUMATIC FEVER*

Definition: acute rheumatism with a tendency to valvular heart-disease.

NATAL: Mars and Saturn afflicted, especially afflicting each other.

PROGRESSED: an aspect, especially an affliction, involving either Saturn or Mars.

112. *RHEUMATISM*

Definition: a disease with fever, pain, inflammation, and swelling of the joints.

NATAL: Mars and Saturn afflicted, especially afflicting each other.

PROGRESSED: an aspect, especially an affliction, involving either Saturn or Mars.

113. *RHINITIS*

Definition: inflammation of the nasal mucous membrane.

NATAL: an upper-octave planet prominent and Jupiter and Mars afflicted.

PROGRESSED: an aspect involving Mars; an aspect involving Jupiter; and an aspect involving an upper-octave planet.

114. *RICKETS* (*Rachitis*)

Definition: a constitutional disease of childhood marked by increased cell-growth of the bones, deficiency of earthy matter, deformities and changes in the liver and spleen.

NATAL: an upper-octave planet prominent and Saturn afflicted.

PROGRESSED: an aspect involving an upper-octave planet, and an aspect involving Saturn with severe Rallying Forces.

115. *RINGWORM*

Definition: a circling skin-disease, from fungi, with patches of distinct vesicles (small blisters).

NATAL: Mars and Venus somewhat afflicted.

PROGRESSED: an aspect involving Mars and an aspect involving Venus.

116. *SCABIES*

Definition: the itch, a contagious parasitic skin-disease.

NATAL: Mars and Venus somewhat afflicted.

PROGRESSED: an aspect involving Mars and an aspect involving Venus at the same time there are severe Rallying Forces.

117. *SCIATICA*

Definition: neuralgia of the sciatic nerve in the inferior part of the hip-bone.

NATAL: Mercury or Uranus afflicted; and a planet afflicted in Sagittarius or less commonly Jupiter afflicted.

PROGRESSED: an affliction or an aspect involving Mercury or Uranus, at the same time there are severe Rallying Forces; and an aspect involving a planet in Sagittarius or, less commonly, involving Jupiter.

118. *SHINGLES* (*Herpes-Zoster*)

Definition: an acute, inflammatory, painful disease of the skin, consisting of grouped vesicles corresponding in distribution to the course of cutaneous nerves.

NATAL: an upper-octave planet prominent and afflicted; Venus and Mercury afflicted.

PROGRESSED: an aspect involving an upper-octave planet; an aspect involving Venus; and an aspect involving Mercury with heavily active Rallying Forces.

119. *SINUS TROUBLE (Nasal)*

Definition: pertaining to the two irregular cavities in the frontal bone containing air and communicating with the nose by the infundibulum.

NATAL: an upper-octave planet prominent, Saturn or Jupiter afflicted, and commonly a planet in either Aries or Scorpio.

PROGRESSED: an aspect involving an upper-octave planet, an aspect involving Saturn or Jupiter, and an aspect involving the planet in Scorpio or Aries, with severe Rallying Forces.

120. *SLEEPING SICKNESS (Encephalitis lethargica)*

Definition: a peculiar epidemic disease characterized by increasing inclination to sleep (somnolence).

NATAL: Neptune prominent and afflicted.

PROGRESSED: an aspect involving Neptune and an aspect involving Mars.

121. *SMALL-POX (Variola)*

Definition: a specific infectious disease with fever and papular eruption, followed by vesicles, pustules and the production of pits.

NATAL: Mars and Venus prominent and afflicted.

PROGRESSED: an aspect involving Mars and an aspect involving Venus at the same time there are severe Rallying Forces.

122. *SPINAL CURVATURE*

Definition: a bending of the axis of the spine, due to disease or to defective muscular action.

NATAL: Sun and Saturn involved in an affliction or Saturn and a planet in Leo involved in an affliction.

PROGRESSED: an affliction involving the Sun or an affliction involving the Leo planet.

123. *SPINAL MENINGITIS*

Definition: inflammation affecting the membranes of the spinal cord.

NATAL: an upper-octave planet prominent; Sun afflicted; and Mars afflicted or an afflicted planet in Aries.

PROGRESSED: an aspect involving an upper-octave planet, an aspect involving the Sun, and an aspect involving Mars or the Aries planet accompanied by severe Rallying Forces.

124. *SPLEEN TROUBLE*

Definition: pertaining to the oval viscus behind the outer end of the stomach.

NATAL: an affliction to either Sun or Saturn; often Sun and Saturn involved in an affliction.

PROGRESSED: an affliction involving Sun or Saturn, or afflictions involving both.

125. *SPRAIN*

Definition: a violent straining of ligaments.

NATAL: Mars prominent and severely afflicted.

PROGRESSED: an aspect involving Mars at the same time there are severe Rallying Forces.

126. *STUTTERING*

Definition: hesitation and repetition in speaking.

NATAL: Uranus prominent and Mercury afflicted.

PROGRESSED: an aspect involving Mercury with severe Rallying Forces especially involving Uranus.

127. *SUICIDE*

Definition: self-murder.

NATAL: Mars afflicted and a planet in the twelfth house, or the ruler of the twelfth house heavily afflicted.

PROGRESSED: an aspect to the ruler of the twelfth house at the same time there are severe Rallying Forces.

128. *SUN STROKE (Heat Stroke)*

Definition: heat stroke from direct rays of the sun.

NATAL: watery planets, chiefly Neptune or the Moon, prominent and afflicted; often an affliction in Aries.

PROGRESSED: an aspect, usually an affliction, involving the Sun; and an aspect usually an affliction, involving Mars.

129. *SYPHILIS*

Definition: a chronic, infectious, venereal disease, which may also be hereditary, inducing cutaneous and other lesions.

NATAL: Venus or Mars, usually both, afflicted; especially an affliction involving Mars and Venus.

PROGRESSED: an aspect, usually an affliction, involving Mars.

130. *THROMBOSIS*

Definition: formation of a bloodclot in the heart or in a blood-vessel.

NATAL: Venus or Jupiter afflicted, and Pluto, Neptune or Saturn prominent and afflicted.

PROGRESSED: an affliction involving Venus or Jupiter at the same time there is an aspect involving Pluto, Neptune or Saturn, and an aspect involving Mars.

131. *TICK FEVER*

Definition: infectious disease caused by a tick-transmitted protozoan parasite to man and animals.

NATAL: Mars prominent and Neptune prominent and afflicted.

PROGRESSED: an aspect involving Mars and an aspect involving Neptune, with severe Rallying Forces.

132. *TOOTH TROUBLE*

Definition: pertaining to the organs of mastication.

NATAL: Saturn prominent and afflicted, and an upper-octave planet prominent.

PROGRESSED: an aspect involving Saturn and an aspect involving Mercury or an upper-octave planet at the same time severe Rallying Forces are present.

133. *TRENCH MOUTH (Gingivitis)*

Definition: inflammation of the gums.

NATAL: Saturn and Mars prominent and afflicted.

PROGRESSED: an aspect involving Saturn and an aspect involving Mars, accompanied by severe Rallying Forces.

134. *TUMOR (Benign)*

Definition: a growth or swelling not giving rise to metastasis (change in the seat of the disease) nor recurring after removal.

NATAL: Jupiter, Saturn, Neptune and Moon prominent and often afflicted.

PROGRESSED: an aspect involving Jupiter and an aspect involving Saturn.

135. *ULCER*

Definition: suppuration upon a free surface; an open sore.

NATAL: Mars and Uranus afflicted and Saturn, Neptune or Pluto prominent and usually afflicted.

PROGRESSED: an aspect involving Mars; an aspect involving Uranus; and an aspect involving Saturn, Neptune or Pluto, with severe Rallying Forces present.

136. *UNDULANT FEVER (Mediterranean Fever)*

Definition: a specific febrile disease of the Mediterranean coast, caused by Alcaligenes abortus.

NATAL: Mars and Moon prominent and afflicted.

PROGRESSED: an aspect involving Mars and an aspect involving the Moon, with serious Rallying Forces.

137. *UTERUS TROUBLE*

Definition: pertaining to the hollow female organ of gestation.

NATAL: a planet afflicted in Scorpio, or an affliction involving Mars or Venus, especially if the Moon is afflicted.

PROGRESSED: an aspect, usually an affliction, involving the Scorpio planet or involving Mars or Venus, especially if the Moon is also involved in an aspect.

138. *VARICOSE VEINS*

Definition: swollen, knotted blood-vessels.

NATAL: Venus afflicted.

PROGRESSED: an aspect involving Venus at the same time there are severe Rallying Forces, especially involving Mars.

139. *VITUS'S DANCE, ST. (Chorea)*

Definition: involuntary muscular twitchings.

NATAL: an aspect between Mercury and Uranus, usually an affliction.

PROGRESSED: an aspect involving Uranus at the same time severe Rallying Forces are present.

140. *WHOOPING COUGH (Pertussis)*

Definition: a contagious, convulsive cough.

NATAL: Pluto, Neptune or Saturn prominent, and Mercury or a planet in Gemini afflicted.

PROGRESSED: an aspect involving Mercury or the Gemini planet, and usually an aspect involving Saturn, Neptune or Pluto.

141. *YELLOW FEVER*

Definition: an epidemic disease with high fever, jaundice, black vomit, etc.

NATAL: Mars afflicted and prominent; and Saturn, Neptune or Pluto prominent.

PROGRESSED: an affliction involving Mars at the same time Saturn, Neptune or Pluto are involved in an aspect.

V
CALCULATION OF THE CHART

No matter what branch of astrology it is cast for, the chart is erected by the same rules. Volunteer research assistants follow the standard method of chart erection, using Church of Light Student Chart Blanks. These blanks, designed to facilitate each essential step in chart erection, present a line to be filled in following instructions as to the entry made upon it, serving as a necessary check against overlooking some required calculation.

This particular form of chart blank has proven invaluable, because no matter how long ago the chart was erected, a glance reveals what tables were used and the basic math of its construction. When just the wheel is kept for reference, a great deal of laborious work is involved in checking for accuracy.

28 lines appear on the chart blank in order to handle any contingency in chart erection; however, sometimes a few lines—not applying to the particular chart under construction—are left blank. The birth-charts illustrated in this book have been calculated on such blanks—as are all the charts in the research file. To understand the following instructions (based on Greenwich Mean Noon Ephemeris), it is suggested that the correct number be inserted before each of the 28 lines to the left of the illustrated charts.

Line 1: Write in name of person, corporation, city, nation, planetary cycle, horary question or file number of case history for which chart is erected.

Line 2: Write in month, day of month and year of birth.

Line 3: Write in name of birth place.

Line 4: Write in latitude of birth place.

Line 5: Write in longitude of birth place.

(Lines 6 through 9 are not used by research assistants, because all birth times on case histories are given as Local Mean Time. Calculation of LMT is completed before the case history is filed—first the kind of time for the specific year and place is checked, then the time is converted to LMT. See Table III, Local Mean Time Correction.)

Line 6: Unless another kind of time than local mean time was used in timing birth, leave blank.

Line 7: If another time than standard or local mean time was used in timing birth, enter difference from standard on this line.

Line 8: If birth was timed by standard time, write it in. If birth time was other than standard or local mean time, add or subtract correction given on line 7.

Line 9: If birth was timed by local mean time, leave blank. Otherwise, write in correction of standard time to local mean time. Consult Table III, Correction of Standard to Local Mean Time, or find the difference in degrees and minutes between the longitude of birth and the longitude of the Standard Time Meridian of the time zone within which birth took place. Multiply the degrees so found by 4, calling the product minutes of time. Multiply the minutes so found by 4, calling the product seconds of time. Write the result on line 9. If birth longitude is east of standard meridian place a plus sign before this time interval; if west, a minus sign.

Line 10: If the birth was timed by local mean time, write it in. Otherwise, add or subtract, as sign indicates, line 9 to or from line 8, writing the sum or difference on line 10. In the research charts local mean time is expressed in 2400 hours; viz., 01:30:32 p.m. appears as 13:30:32 LMT. However, the A.M. or P.M. may be crossed off the blank to indicate if the birth was before or after noon. (When the time is from 12 noon to 12:59 p.m. it is more correct to write it as 0 hours and so many minutes p.m.; for example, 12:15 p.m. is more correctly given as 00:15 p.m., or 15 minutes after noon.)

Line 11: Represents 12:00 noon. Sometimes it is convenient to write in 11:59:60 (for purposes of subtraction) which equals noon, or 12 hours, on the day of birth.

Line 12: Write in local mean time of birth (from line 10).

Line 13: If LMT is A.M., subtract line 12 from line 11, preceding the hours, minutes and seconds so found by a minus sign. If a zero hour ephemeris is used (instead of noon), add 12 hours (the time interval from midnight to noon) to the LMT before entering it on this line.

Line 14: From the birthday ephemeris, copy the hours, minutes and seconds of sidereal time there given.

Line 15: If line 13, and therefore line 17 is minus and greater than line 14, in order to subtract line 17 from line 14, 24 hours should be written on line 15. Otherwise, line 15 is left blank.

Line 16: Left blank unless it becomes necessary to write 24 hours on line 15. In these instances, the hours on line 15 should be added to the hours, minutes and seconds of line 14 and their sum written on line 16.

Line 17: Copy the hours, minutes and seconds of line 13 preceded by their plus or minus sign, on line 17.

Line 18: Add or subtract, as the plus or minus sign indicates, line 17 to or from line 14 (or from line 16 if an entry on it was necessary). Write the hours, minutes and seconds so found on line 18.

Before making an entry on line 19, it is first necessary to ascertain line 26. For now, lines 19 and 20 are skipped.

Line 21: Write in the Local Mean Time of birth.

Line 22: The longitude of birth should be converted into hours, minutes and seconds of time by multiplying the degrees by 4 and calling the product minutes, and multiplying the minutes by 4 and calling the product seconds. If the birth was in east longitude precede the hours, minutes and seconds so found by a minus sign; if in west longitude, by a plus sign.

Line 23: As indicated by the plus or minus sign, add line 22 to, or subtract from, line 21. Designate A.M. or P.M.

Line 24: Represents 12:00 noon. Sometimes it is convenient to write in 11:59:60 (for purposes of subtraction) which equals noon, or 12 hours, on the day of birth.

Line 25: If line 23 is P.M., line 25 is left blank. If line 23 is A.M., copy on line 25.

Line 26: If line 23 is P.M., its hours, minutes and seconds are copied on line 26 preceded by a plus sign. If line 23 is A.M., it is subtracted from line 24 and the difference in hours, minutes and seconds written on line 26 preceded by a minus sign. The Plus or Minus Equivalent Greenwich Mean Time Interval of line 26 is the interval from which the sign, degree and minute occupied by each of the ten planets is calculated.

Now we go back to line 19.

Line 19: The hours, minutes and seconds of line 26 are corrected at the rate of 9.86 seconds per hour. (See Table IV, Correction of Mean to Sidereal Time.) Precede sum so found by a minus sign if line 26 is minus; by a plus sign, if line 26 is plus.

Line 20: If line 19 is minus, subtract from line 18. If line 19 is plus, add to line 18. Write the hours, minutes and seconds so found on line 20. This is the Sidereal Time of Birth from which, with the aid of a table of houses for the birth latitude, the sign, degree and minute on each house cusp of the chart may be calculated. If the chart is erected for a place in south latitude, 12 hours should be added to line 20, using the degrees thus found, but placing opposite signs on the house cusps.

In the space after line 26, note which ephemeris and which table of houses was used to erect chart. There are just two more lines.

Line 27: Copy from a table of diurnal proportional logarithms the number corresponding to the hours and minutes on line 26. See Table V.

Line 28: If no progressions are to be figured, line 28 may left blank. In the research charts, we always enter the Limiting Date; then it is ready when needed. The hours of line 26 are divided by 2 and called months, and the minutes are divided by 4 and called days. (See Table VIII, Correction of Time Interval to Calendar Interval. If line 26 is minus, the months and days so found are added to the year, month and day of birth. However, if line 26 is plus, they are subtracted from the birthday. The calendar date so found, including the year, is written on line 28. This Limiting Date is the starting point in calculating the major progressed positions on any calendar date, and in calculating the calendar date on which any major progressed aspect is perfect.

The basic math of chart erection is now completed. The reasons why these basic calculations, as well as those to follow, must be taken to produce a precise birth-chart for a specific date, hour and place are given, with examples, in the book "Horary Astrology", by C. C. Zain.

CALCULATION OF PRECISE HOUSE CUSPS

The following short-cut methods have been adopted, using a table of houses for birth latitude, in erecting charts for the Northern Hemisphere. See line 20 in "Calculation of the Chart" for instructions regarding south latitude charts.

A. Difference in minutes and seconds between the nearest and next nearest Sidereal Time to that of birth (line 20) in the Table.

B. Difference in minutes and seconds between the True Sidereal Time of birth and the nearest Sidereal Time in the Table to that of birth.

C. Difference between the house cusps corresponding to the nearest and next nearest Sidereal Time in the Table to that of birth.

D. Correction to be made.

PROBLEM BY LOGS: (Convert A, B and C to Logs.) B minus A equals Basic Log (to be used on all cusp corrections) plus C equals D. (Convert Log of D into degrees and minutes of longitude.)

TO CORRECT: Add D to cusp given in Table if nearest Sidereal Time is smaller than True Sidereal Time; if larger, subtract D.

If the Table of Houses was for the exact latitude of birth,

no further cusp correction is necessary. If not, then the following problem will correct cusps to precise latitude of birth. The Midheaven (and thus the 4th cusp) remains the same in any latitude; but the correction must be applied to each other cusp.

CORRECTION FOR PRECISE LATITUDE

A. Difference between nearest and next nearest Latitude in Table to that of birth.

B. Difference between nearest Latitude in Table and Birth Latitude.

C. Difference between cusps corresponding to nearest and next nearest Latitude in Table to that of birth.

D. Correction to be made.

PROBLEM BY LOGS: (Convert A, B, C to Logs.) B minus A equals Basic Log (to be used on all cusp corrections) plus C equals D. (Convert Log of D into degrees and minutes of longitude.)

TO CORRECT: Add D to cusps determined in problem above if the nearest Latitude is smaller than the True; if larger, subtract D.

By using these two problems, the precise house cusps can be ascertained and placed on the wheel.

CORRECTION FOR PLANETARY POSITION

The constant log used to correct planets is found on line 27. To correct: add log of planet's daily motion (same problem used to determine planet's zodiacal position and by declination) to the constant log; the log so found is converted into time. This is the correction to be made from the noon position given in the ephemeris. If the EGMT Interval entered on line 26 is plus, add the correction so found to planet's position on the noon of birthday; if line 26 is minus, subtract correction.

As only nine planetary positions are given in most ephemerides, Pluto's — the tenth planet — position is taken from "The Influence of Pluto", by Elbert Benjamine. This particular ephemeris gives the Pluto position every 30 days, thus the following problem is applicable to that ephemeris only.

CORRECTION FOR PLUTO'S POSITION

A. Number of days in ephemeris between nearest and next nearest day to birthday.

B.* Pluto's travel in that time.

C. Interval of days between nearest Pluto date in ephemeris and birthday.

D. Correction to be made.

PROBLEM BY LOGS: (Convert A, B and C to Logs.) Find difference between A and B, plus C equals D. (Convert Log of D to degrees and minutes.)

TO CORRECT: Add D to position on nearest date given in ephemeris if date is smaller than birthday; if larger, subtract.

This problem is used for both the zodiacal position of Pluto and its position in declination.

At this point all cusps and planets have been corrected and entered in their appropriate places on the wheel, and the declinations of the ten planets listed. Only information missing is the Midheaven and Ascendant declinations. Both points are called angles, and the same rule applies.

DECLINATION OF THE ANGLES

A. Difference between the longitude of the Sun in an ephemeris nearest and next nearest to the sign, degree and minute on the Angle.

B. Difference between nearest Sun longitude and longitude of Angle.

C. Difference between declinations corresponding to nearest and next nearest position of Sun in the ephemeris used in A.

D. Correction to be made.

PROBLEM BY LOGS: (Convert A, B and C to Logs.) B minus A plus C equals D. (Convert D to degrees and minutes.)

TO CORRECT: Add D to nearest declination in ephemeris if longitude of the Sun is smaller than longitude of Angle; if larger, subtract. (See Table VI.)

If the rules are followed, the resulting chart is as precise as available birth data and tables will allow.

*If motion is over 30′, use proportion rather than logs.

VI

Research Department Tables

Table I
LONGITUDE MEASURE

60	seconds	(″)	equal 1 minute,	marked ′
60	minutes	(′)	equal 1 degree,	marked °
30	degrees	(°)	equal 1 sign,	marked S
12	signs	(S)	equal 1 zodiac	
360	degrees	(°)	equal 1 circle	

Table II
TIME MEASURE

60	seconds	(sec.)	equal 1 minute,	marked m
60	minutes	(min.)	equal 1 hour,	marked h
24	hours	(hr.)	equal 1 day,	marked D
7	days	(D)	equal 1 week	
365¼	days		equal 1 year	
52+	weeks		equal 1 year	

Table III
CORRECTION OF STANDARD TO LOCAL MEAN TIME

MINUTES				
°	1	2	3	″
	15	30	45	00
1	16	31	46	04
2	17	32	47	08
3	18	33	48	12
4	19	34	49	16
5	20	35	50	20
6	21	36	51	24
7	22	37	52	28
8	23	38	53	32
9	24	39	54	36
10	25	40	55	40
11	26	41	56	44
12	27	42	57	48
13	28	43	58	52
14	29	44	59	56

INSTRUCTIONS: In chart erection to correct standard clock time to local mean time, the interval of degrees and minutes between longitude of birth and the longitude of the standard meridian where the birth took place is converted into time by multiplying the degrees so found by 4, calling the product minutes of time; and multiplying the minutes so found by 4, calling the product seconds of time. The table performs this multiplication.

EXAMPLE: Find the LMT correction for a longitude variation of 3° and 34′. In extreme left column find 3 under °; follow horizontal line to extreme right column to find answer, 12 under ″. Look to body of table for 34 (middle column); answer is found by projecting a right angle—at head of this column is 2′ and at extreme right is 16″, which added to the 12′ found above gives a total correction of 14′ and 16″.

TABLE IV

CORRECTION OF MEAN TO SIDEREAL TIME

Mean Time	Correction		Mean Time	Correction	Mean Time	Correction	Mean Time	Correction	Mean Time	Correction
H.	M.	s.	M.	s.	M.	s.	S.	s.	S.	s.
1	0	9.86	1	0.16	31	5.09	1	.00	31	.09
2	0	19.71	2	0.33	32	5.26	2	.00	32	.09
3	0	29.57	3	0.49	33	5.42	3	.01	33	.09
4	0	39.43	4	0.66	34	5.58	4	.01	34	.09
5	0	49.28	5	0.82	35	5.75	5	.01	35	.10
6	0	59.14	6	0.99	36	5.91	6	.02	36	.10
7	1	9.00	7	1.15	37	6.08	7	.02	37	.10
8	1	18.85	8	1.31	38	6.24	8	.02	38	.10
9	1	28.71	9	1.48	39	6.41	9	.02	39	.11
10	1	38.57	10	1.64	40	6.57	10	.03	40	.11
11	1	48.42	11	1.81	41	6.73	11	.03	41	.11
12	1	58.28	12	1.97	42	6.90	12	.03	42	.11
13	2	8.13	13	2.14	43	7.06	13	.04	43	.12
14	2	17.99	14	2.30	44	7.23	14	.04	44	.12
15	2	27.85	15	2.46	45	7.39	15	.04	45	.12
16	2	37.70	16	2.63	46	7.56	16	.04	46	.13
17	2	47.56	17	2.79	47	7.72	17	.05	47	.13
18	2	57.42	18	2.96	48	7.88	18	.05	48	.13
19	3	7.27	19	3.12	49	8.05	19	.05	49	.13
20	3	17.13	20	3.28	50	8.21	20	.05	50	.14
21	3	26.99	21	3.45	51	8.38	21	.06	51	.14
22	3	36.84	22	3.61	52	8.54	22	.06	52	.14
23	3	46.70	23	3.78	53	8.71	23	.06	53	.15
			24	3.94	54	8.87	24	.07	54	.15
			25	4.11	55	9.03	25	.07	55	.15
			26	4.27	56	9.20	26	.07	56	.15
			27	4.43	57	9.36	27	.07	57	.16
			28	4.60	58	9.53	28	.08	58	.16
			29	4.76	59	9.69	29	.08	59	.16
			30	4.93	60	9.86	30	.08	60	.16

INSTRUCTIONS: In chart erection, the EGMTI is corrected at the rate of 9.86s per hour to convert uncorrected sidereal time into true sidereal time of birth. The table performs this multiplication, and can be used for any problems involving the conversion of mean to sidereal time.

EXAMPLE: Find correction for EGMTI of 7h, 28m and 33s.

7h (first column) equals	1m 9.00s	
28m (second column) equals	4.60s	
33s (last column) equals	.09s	add
	1m 13.69s, or 1m, 14s correction	

TABLE V
DIURNAL PROPORTIONAL LOGARITHMS

Min.	Degree or Hour											
	0	**1**	**2**	**3**	**4**	**5**	**6**	**7**	**8**	**9**	**10**	**11**
0	Infinite	1.38021	1.07918	.90309	.77815	.68124	.60206	.53511	.47712	.42597	.38021	.33882
1	3.15836	1.37303	1.07558	.90069	.77635	.67980	.60086	.53408	.47622	.42517	.37949	.33816
2	2.85733	1.36597	1.07200	.89829	.77455	.67836	.59966	.53305	.47532	.42436	.37877	.33751
3	2.68124	1.35902	1.06846	.89591	.77276	.67692	.59846	.53202	.47442	.42356	.37805	.33685
4	2.55630	1.35218	1.06494	.89355	.77097	.67549	.59726	.53100	.47352	.42276	.37733	.33620
5	2.45939	1.34545	1.06145	.89119	.76920	.67406	.59607	.52997	.47262	.42197	.37661	.33554
6	2.38021	1.33882	1.05799	.88885	.76743	.67264	.59488	.52895	.47173	.42117	.37589	.33489
7	2.31327	1.33229	1.05456	.88652	.76567	.67123	.59370	.52794	.47083	.42038	.37518	.33424
8	2.25527	1.32585	1.05115	.88421	.76391	.66981	.59252	.52692	.46994	.41958	.37446	.33359
9	2.20412	1.31951	1.04777	.88190	.76216	.66841	.59134	.52591	.46905	.41879	.37375	.33294
10	2.15836	1.31327	1.04442	.87961	.76042	.66700	.59016	.52490	.46817	.41800	.37303	.33229
11	2.11697	1.30711	1.04109	.87733	.75869	.66560	.58899	.52389	.46728	.41721	.37232	.33164
12	2.07918	1.30103	1.03779	.87506	.75696	.66421	.58782	.52288	.46640	.41642	.37161	.33099
13	2.04442	1.29504	1.03451	.87281	.75524	.66282	.58665	.52188	.46552	.41564	.37090	.33035
14	2.01224	1.28913	1.03126	.87056	.75353	.66143	.58549	.52087	.46464	.41485	.37020	.32970
15	1.98227	1.28330	1.02803	.86833	.75182	.66005	.58433	.51987	.46376	.41407	.36949	.32906
16	1.95424	1.27755	1.02482	.86611	.75012	.65868	.58318	.51888	.46288	.41329	.36878	.32842
17	1.92791	1.27187	1.02164	.86390	.74843	.65730	.58202	.51788	.46201	.41251	.36808	.32778
18	1.90309	1.26627	1.01848	.86170	.74674	.65594	.58087	.51689	.46113	.41173	.36738	.32713
19	1.87961	1.26074	1.01535	.85951	.74506	.65457	.57972	.51590	.46026	.41095	.36667	.32649
20	1.85733	1.25527	1.01224	.85733	.74339	.65321	.57858	.51491	.45939	.41018	.36597	.32585
21	1.83614	1.24988	1.00914	.85517	.74172	.65186	.57744	.51393	.45853	.40940	.36527	.32522
22	1.81594	1.24455	1.00608	.85301	.74006	.65051	.57630	.51294	.45766	.40863	.36457	.32458
23	1.79664	1.23929	1.00303	.85087	.73841	.64916	.57516	.51196	.45680	.40786	.36388	.32394
24	1.77815	1.23408	1.00000	.84873	.73676	.64782	.57403	.51098	.45593	.40709	.36318	.32331
25	1.76042	1.22894	0.99700	.84661	.73512	.64648	.57290	.51000	.45507	.40632	.36248	.32267
26	1.74339	1.22387	0.99401	.84450	.73348	.64514	.57178	.50903	.45421	.40555	.36179	.32204
27	1.72700	1.21884	0.99105	.84239	.73185	.64382	.57065	.50806	.45336	.40478	.36110	.32141
28	1.71121	1.21388	0.98810	.84030	.73023	.64249	.56953	.50709	.45250	.40402	.36040	.32078
29	1.69597	1.20897	0.98518	.83822	.72861	.64117	.56841	.50612	.45165	.40325	.35971	.32014
30	1.68124	1.20412	0.98227	.83614	.72700	.63985	.56730	.50515	.45079	.40249	.35902	.31951
31	1.66700	1.19932	0.97939	.83408	.72539	.63854	.56619	.50419	.44994	.40173	.35833	.31889
32	1.65321	1.19458	0.97652	.83203	.72379	.63723	.56508	.50323	.44909	.40097	.35765	.31826
33	1.63985	1.18988	0.97367	.82998	.72220	.63592	.56397	.50227	.44825	.40021	.35696	.31763
34	1.62688	1.18524	0.97084	.82795	.72061	.63462	.56287	.50131	.44740	.39945	.35627	.31700
35	1.61430	1.18064	0.96803	.82593	.71903	.63332	.56177	.50035	.44656	.39870	.35559	.31638
36	1.60206	1.17609	0.96524	.82391	.71745	.63202	.56067	.49940	.44571	.39794	.35491	.31575
37	1.59016	1.17159	0.96246	.82190	.71588	.63073	.55957	.49845	.44487	.39719	.35422	.31513
38	1.57858	1.16714	0.95971	.81991	.71432	.62945	.55848	.49750	.44403	.39644	.35354	.31451
39	1.56730	1.16273	0.95697	.81792	.71276	.62816	.55739	.49655	.44320	.39569	.35286	.31389
40	1.55630	1.15836	0.95424	.81594	.71121	.62688	.55630	.49560	.44236	.39494	.35218	.31327
41	1.54558	1.15404	0.95154	.81397	.70966	.62561	.55522	.49466	.44153	.39419	.35151	.31265
42	1.53511	1.14976	0.94885	.81201	.70811	.62434	.55414	.49372	.44069	.39344	.35083	.31203
43	1.52490	1.14554	0.94618	.81006	.70658	.62307	.55306	.49278	.43986	.39270	.35015	.31141
44	1.51491	1.14133	0.94352	.80812	.70505	.62181	.55198	.49185	.43903	.39195	.34948	.31079
45	1.50515	1.13717	0.94088	.80618	.70352	.62054	.55091	.49091	.43820	.39121	.34880	.31017
46	1.49561	1.13306	0.93826	.80426	.70200	.61929	.54984	.48998	.43738	.39047	.34813	.30956
47	1.48627	1.12898	0.93565	.80234	.70048	.61803	.54877	.48905	.43655	.38973	.34746	.30894
48	1.47712	1.12494	0.93305	.80043	.69897	.61678	.54770	.48812	.43573	.38899	.34679	.30833
49	1.46817	1.12094	0.93048	.79853	.69747	.61554	.54664	.48719	.43491	.38825	.34612	.30772
50	1.45939	1.11697	0.92791	.79664	.69597	.61430	.54558	.48627	.43409	.38751	.34545	.30710
51	1.45079	1.11304	0.92537	.79475	.69447	.61306	.54452	.48534	.43327	.38678	.34478	.30649
52	1.44236	1.10915	0.92284	.79288	.69298	.61182	.54347	.48442	.43245	.38604	.34412	.30588
53	1.43409	1.10529	0.92032	.79101	.69150	.61059	.54241	.48350	.43164	.38531	.34345	.30527
54	1.42597	1.10146	0.91781	.78915	.69002	.60936	.54136	.48259	.43082	.38458	.34279	.30467
55	1.41800	1.09767	0.91533	.78730	.68854	.60814	.54032	.48167	.43001	.38385	.34212	.30406
56	1.41018	1.09391	0.91285	.78545	.68707	.60691	.53927	.48076	.42920	.38312	.34146	.30345
57	1.40249	1.09018	0.91039	.78362	.68561	.60570	.53823	.47985	.42839	.38239	.34080	.30284
58	1.39494	1.08648	0.90794	.78179	.68415	.60448	.53719	.47894	.42758	.38166	.34014	.30224
59	1.38751	1.08282	0.90551	.77997	.68269	.60327	.53615	.47803	.42677	.38094	.33948	.30163

INSTRUCTIONS: This table may be used to find the diurnal log for intervals of degrees or hours and minutes, as well as intervals of minutes and seconds. EXAMPLES: Find log for 7h 13m: Look for 7 at top of table, go down that column until reaching 13, designated on sides of table; answer is .52188.

DIURNAL PROPORTIONAL LOGARITHMS (Continued)

Min.	Degree or Hour											
	12	**13**	**14**	**15**	**16**	**17**	**18**	**19**	**20**	**21**	**22**	**23**
0	.30103	.26627	.23408	.20412	.17609	.14976	.12494	.10146	.07918	.05799	.03779	.01848
1	.30043	.26571	.23357	.20364	.17564	.14934	.12454	.10108	.07882	.05765	.03746	.01817
2	.29983	.26516	.23305	.20316	.17519	.14891	.12414	.10070	.07846	.05730	.03713	.01786
3	.29923	.26460	.23254	.20268	.17474	.14849	.12374	.10032	.07810	.05696	.03680	.01754
4	.29863	.26405	.23202	.20220	.17429	.14806	.12333	.09994	.07774	.05662	.03648	.01723
5	.29803	.26349	.23151	.20172	.17384	.14764	.12293	.09956	.07738	.05627	.03615	.01691
6	.29743	.26294	.23099	.20124	.17339	.14722	.12253	.09918	.07702	.05593	.03582	.01660
7	.29683	.26239	.23048	.20076	.17294	.14679	.12213	.09880	.07666	.05559	.03549	.01629
8	.29623	.26184	.22997	.20028	.17249	.14637	.12173	.09842	.07630	.05524	.03517	.01597
9	.29564	.26129	.22946	.19980	.17204	.14595	.12134	.09804	.07594	.05490	.03484	.01566
10	.29504	.26074	.22894	.19932	.17159	.14553	.12094	.09767	.07558	.05456	.03451	.01535
11	.29445	.26019	.22843	.19885	.17114	.14511	.12054	.09729	.07522	.05422	.03419	.01504
12	.29385	.25964	.22792	.19837	.17070	.14468	.12014	.09691	.07486	.05388	.03386	.01472
13	.29326	.25909	.22741	.19789	.17025	.14426	.11974	.09653	.07450	.05354	.03353	.01441
14	.29267	.25854	.22691	.19742	.16980	.14384	.11935	.09616	.07415	.05319	.03321	.01410
15	.29208	.25800	.22640	.19694	.16936	.14342	.11895	.09578	.07379	.05285	.03288	.01379
16	.29149	.25745	.22589	.19647	.16891	.14300	.11855	.09541	.07343	.05251	.03256	.01348
17	.29090	.25691	.22538	.19599	.16847	.14258	.11816	.09503	.07307	.05217	.03223	.01317
18	.29031	.25636	.22488	.19552	.16802	.14217	.11776	.09466	.07272	.05183	.03191	.01286
19	.28972	.25582	.22437	.19505	.16758	.14175	.11737	.09428	.07236	.05149	.03158	.01255
20	.28913	.25527	.22387	.19458	.16714	.14133	.11697	.09391	.07200	.05115	.03126	.01224
21	.28855	.25473	.22336	.19410	.16669	.14091	.11658	.09353	.07165	.05081	.03093	.01193
22	.28796	.25419	.22286	.19363	.16625	.14050	.11618	.09316	.07129	.05048	.03061	.01162
23	.28737	.25365	.22235	.19316	.16581	.14008	.11579	.09278	.07094	.05014	.03029	.01131
24	.28679	.25311	.22185	.19269	.16537	.13966	.11539	.09241	.07058	.04980	.02996	.01100
25	.28621	.25257	.22135	.19222	.16493	.13925	.11500	.09204	.07023	.04946	.02964	.01069
26	.28562	.25203	.22085	.19175	.16449	.13883	.11461	.09167	.06987	.04912	.02932	.01038
27	.28504	.25149	.22034	.19128	.16405	.13842	.11422	.09129	.06952	.04879	.02900	.01007
28	.28446	.25095	.21984	.19082	.16361	.13800	.11382	.09092	.06917	.04845	.02867	.00976
29	.28388	.25042	.21934	.19035	.16317	.13759	.11343	.09055	.06881	.04811	.02835	.00945
30	.28330	.24988	.21884	.18988	.16273	.13717	.11304	.09018	.06846	.04777	.02803	.00914
31	.28272	.24934	.21835	.18941	.16229	.13676	.11265	.08981	.06811	.04744	.02771	.00884
32	.28215	.24881	.21785	.18895	.16185	.13635	.11226	.08944	.06775	.04710	.02739	.00853
33	.28157	.24827	.21735	.18848	.16141	.13594	.11187	.08907	.06740	.04677	.02707	.00822
34	.28099	.24774	.21685	.18802	.16098	.13552	.11148	.08870	.06705	.04643	.02675	.00791
35	.28042	.24721	.21636	.18755	.16054	.13511	.11109	.08833	.06670	.04609	.02642	.00761
36	.27984	.24667	.21586	.18709	.16010	.13470	.11070	.08796	.06634	.04576	.02610	.00730
37	.27927	.24614	.21536	.18662	.15967	.13429	.11031	.08759	.06599	.04542	.02578	.00699
38	.27869	.24561	.21487	.18616	.15923	.13388	.10992	.08722	.06564	.04509	.02546	.00669
39	.27812	.24508	.21437	.18570	.15880	.13347	.10953	.08685	.06529	.04475	.02514	.00638
40	.27755	.24455	.21388	.18524	.15836	.13306	.10915	.08648	.06494	.04442	.02482	.00608
41	.27698	.24402	.21339	.18477	.15793	.13265	.10876	.08611	.06459	.04409	.02451	.00577
42	.27641	.24349	.21290	.18431	.15750	.13224	.10837	.08575	.06424	.04375	.02419	.00546
43	.27584	.24296	.21240	.18385	.15706	.13183	.10798	.08538	.06389	.04342	.02387	.00516
44	.27527	.24244	.21191	.18339	.15663	.13142	.10760	.08501	.06354	.04309	.02355	.00485
45	.27470	.24191	.21142	.18293	.15620	.13101	.10721	.08465	.06319	.04275	.02323	.00455
46	.27413	.24138	.21093	.18247	.15577	.13061	.10683	.08428	.06285	.04242	.02291	.00424
47	.27357	.24086	.21044	.18201	.15533	.13020	.10644	.08391	.06250	.04209	.02260	.00394
48	.27300	.24033	.20995	.18156	.15490	.12979	.10605	.08355	.06215	.04176	.02228	.00364
49	.27244	.23981	.20946	.18110	.15447	.12939	.10567	.08318	.06180	.04142	.02196	.00333
50	.27187	.23929	.20897	.18064	.15404	.12898	.10529	.08282	.06145	.04109	.02164	.00303
51	.27132	.23876	.20849	.18018	.15361	.12857	.10490	.08245	.06111	.04076	.02133	.00272
52	.27075	.23824	.20800	.17973	.15318	.12817	.10452	.08209	.06076	.04043	.02101	.00242
53	.27018	.23772	.20751	.17927	.15275	.12776	.10413	.08172	.06041	.04010	.02069	.00212
54	.26962	.23720	.20703	.17882	.15233	.12736	.10375	.08136	.06007	.03977	.02038	.00181
55	.26906	.23668	.20654	.17836	.15190	.12696	.10337	.08100	.05972	.03944	.02006	.00151
56	.26850	.23616	.20606	.17791	.15147	.12655	.10299	.08063	.05937	.03911	.01975	.00121
57	.26794	.23564	.20557	.17745	.15104	.12615	.10260	.08027	.05903	.03878	.01943	.00091
58	.26738	.23512	.20509	.17700	.15062	.12574	.10222	.07991	.05868	.03845	.01911	.00060
59	.26683	.23460	.20460	.17655	.15019	.12534	.10184	.07954	.05834	.03812	.01880	.00030

Find log of 17m 18s: Look to 17 at top, follow down column to 18; answer 14217.

Convert log .74172 into time: Look to body of table (fifth column from left), then read answer at top and side of table: 4h 21m.

TABLE VI

DECLINATION OF ANGLES

♈	N	♍	♉	N	♌	♊	N	♋
♎	S	♓	♏	S	♒	♐	S	♑
Deg.	*Dec.*	*Deg.*	*Deg.*	*Dec.*	*Deg.*	*Deg.*	*Dec.*	*Deg.*
0	0:00		0	11:28		0	20:09	
1	0:23	29	1	11:49	29	1	20:22	29
2	0:47	28	2	12:11	28	2	20:34	28
3	1:12	27	3	12:31	27	3	20:46	27
4	1:36	26	4	12:52	26	4	20:57	26
5	1:59	25	5	13:12	25	5	21:08	25
6	2:23	24	6	13:32	24	6	21:18	24
7	2:47	23	7	13:52	23	7	21:29	23
8	3:11	22	8	14:11	22	8	21:39	22
9	3:34	21	9	14:30	21	9	21:48	21
10	3:58	20	10	14:49	20	10	21:57	20
11	4:22	19	11	15:07	19	11	22:06	19
12	4:45	18	12	15:26	18	12	22:14	18
13	5:08	17	13	15:45	17	13	22:22	17
14	5:32	16	14	16:03	16	14	22:30	16
15	5:55	15	15	16:20	15	15	22:36	15
16	6:18	14	16	16:38	14	16	22:42	14
17	6:41	13	17	16:55	13	17	22:49	13
18	7:04	12	18	17:12	12	18	22:54	12
19	7:27	11	19	17:29	11	19	23:00	11
20	7:49	10	20	17:45	10	20	23:04	10
21	8:11	9	21	18:01	9	21	23:08	9
22	8:34	8	22	18:16	8	22	23:12	8
23	8:56	7	23	18:32	7	23	23:16	7
24	9:19	6	24	18:47	6	24	23:19	6
25	9:41	5	25	19:01	5	25	23:22	5
26	10:02	4	26	19:16	4	26	23:23	4
27	10:25	3	27	19:30	3	27	23:24	3
28	10:46	2	28	19:44	2	28	23:25	2
29	11:07	1	29	19:57	1	29	23:26	1
	11:28	0		20:09	0		23:27	0

INSTRUCTIONS: Declination of M.C. or Asc. is the same as declination given for Sun appearing in same degree and minute. Usual procedure is to look at Sun's declination in the ephemeris when it reaches the same degree and minute of the sign in which the angle appears. By using simple proportion, the above table facilitates the process.

EXAMPLE: Find declination of M.C. 26° Pisces 30′: Look to first column, under Pisces go down to 26 and 27 degrees. Answer must appear between corresponding declinations of 1:36 and 1:12. The intermediate value sought is 26° 30′. 30′ is half a degree, thus half the difference between 1:36 and 1:12 (24′) is 12′. The 12′ must be subtracted from 1:36 (because figures decrease in relation to degrees), and the answer is a declination of 1 S 24. (S, found at the column, equals South.)

TABLE VII

ASPECTS AND THEIR ORBS

Aspect			HOUSES						Aspect		
			CADENT		SUCCEDENT		ANGULAR				
Name	DEG.	SYM.	P	L	P	L	P	L	SYM.	DEG.	Name
SEMI-SEXTILE	30	⚺	1	2	2	3	3	4	⚻	150	INCONJUNCT
SEMI-SQUARE	45	∠	3	4	4	5	5	6	⚼	135	SESQUI-SQUARE
SEXTILE	60	⚹	5	6	6	7	7	8	⚹	60	SEXTILE
SQUARE	90	□	6	8	8	10	10	12	△	120	TRINE
OPPOSITION	180	☍	8	11	10	13	12	15	☌	0	CONJUNCTION

Deg.—degree Sym.—Symbol P—Planets L—Luminaries (Sun, Moon)
M.C. and Asc. are allowed the same orb as planets in angles.
Orb of the Parallel aspect is always 1 degree.

ALWAYS USE THE LARGER ORB.

Effective orb for major, minor or transit progressed aspects — regardless of nature of aspect—is always one degree.

INSTRUCTIONS: Find distance between two planets; check difference against orb of the type aspect and house positions to see if the two planets involved are within the designated orb. Always use larger orb. EXAMPLE: To determine if a seventh-house Uranus in 24° Aries 53′ is in aspect with a Sun in the eleventh house at 5° Virgo 42′:
Find zodiacal difference between two planets:

Sun	6S	5°	42′	
Uranus	1S	24	53	subtract
	4S	10°	49′, or 130° 49′	

Uranus' orb: Look to "Angular" on table, go down in P column to sesqui-square to find orb of 5°.

Sun's orb: Look to "Succedent" (11th house), go down in L column to sesqui-square to find orb of 5°.

In this case, both planets have the same orb—5°. As a sesqui-square aspect is formed at 135°, these two planets appearing 130° 49′ apart are just within the allowable 5° orb of a sesqui-square aspect.

TABLE VIII

CORRECTION OF TIME INTERVAL TO CALENDAR INTERVAL

24 hours equal 1 year

12 hours equal 6 months

2 hours equal 1 month

4 minutes equal 1 day

EXAMPLE: Determine the calendar interval equivalent to 12 hours and 36 minutes: Divide hours by 2 and minutes by 4. Result: 6 months and 9 days.

TABLE IX

MINOR PROGRESSION TIME INTERVAL

Years Age	*Y.*	*Mos.*	*Days*	*Years Age*	*Y.*	*Mos.*	*Days*	*Years Age*	*Y.*	*Mos.*	*Days*	*Years Age*	*Y.*	*Mos.*	*Days*
1	00	00	27	25	01	10	18	49	03	08	03	73	05	05	18
2	00	01	25	26	01	11	15	50	03	09	00	74	05	06	15
3	00	02	22	27	02	00	07	51	03	09	27	75	05	07	13
4	00	03	19	28	02	01	04	52	03	10	25	76	05	08	10
5	00	04	17	29	02	02	02	53	03	11	22	77	05	09	07
6	00	05	14	30	02	02	29	54	04	00	14	78	05	10	04
7	00	06	11	31	02	03	26	55	04	01	12	79	05	11	02
8	00	07	08	32	02	04	24	56	04	02	09	80	05	11	29
9	00	08	06	33	02	05	21	57	04	03	06	81	06	00	21
10	00	09	03	34	02	06	18	58	04	04	03	82	06	01	19
11	00	10	00	35	02	07	16	59	04	05	01	83	06	02	16
12	00	10	28	36	02	08	13	60	04	05	28	84	06	03	13
13	00	11	25	37	02	09	10	61	04	06	25	85	06	04	11
14	01	00	17	38	02	10	07	62	04	07	23	86	06	05	08
15	01	01	15	39	02	11	05	63	04	08	20	87	06	06	05
16	01	02	12	40	02	11	29	64	04	09	17	88	06	07	02
17	01	03	09	41	03	00	24	65	04	10	15	89	06	07	30
18	01	04	06	42	03	01	22	66	04	11	12	90	06	08	27
19	01	05	04	43	03	02	19	67	05	00	04	91	06	09	24
20	01	06	01	44	03	03	16	68	05	01	01	92	06	10	22
21	01	06	28	45	03	04	14	69	05	01	29	93	06	11	19
22	01	07	26	46	03	05	11	70	05	02	26	94	07	00	11
23	01	08	23	47	03	06	08	71	05	03	23	95	07	01	09
24	01	09	20	48	03	07	05	72	05	04	21	96	07	02	06

INSTRUCTIONS: The Mip.D. for any calendar year may be found by counting ahead in the birth ephemeris as many returns of the Moon to the sign, degree and minute it occupies in the natal chart as years of life have elapsed since birth. However, as each calendar year of time is equivalent to 27.3 days of minor progression time in ephemeris, the age can be multiplied by that figure to get the Mip.D. The above table facilitates the process. To the year, month and day of birth, add the elapsed years, months and days of minor progressed time interval to get the Mip.D.

EXAMPLE: Find 1956 Mip.D for native born July 4, 1926:

Calendar year	1956	
Year of birth	1926	subtract
	30	years old in 1956

Look up 30 years in table (second column) to find minor progressed time interval of 2 years, 2 months and 29 days, which is then added to birthday.

	Years	Mo.	Day	
Birth	1926	7	4	
Interval	2	2	29	add
	1928	10	03	

Thus Oct. 3, 1928 in ephemeris is the 1956 Mip.D. for native born July 4, 1926.

TABLE X

MINUTES EXPRESSED AS DECIMALS OF A DEGREE

M.	*Dec.*	*M.*	*Dec.*	*M.*	*Dec.*	*M.*	*Dec.*
1	.02	16	.27	31	.52	46	.77
2	.03	17	.28	32	.53	47	.78
3	.05	18	.30	33	.55	48	.80
4	.07	19	.32	34	.57	49	.82
5	.08	20	.33	35	.58	50	.83
6	.10	21	.35	36	.60	51	.85
7	.12	22	.37	37	.62	52	.87
8	.13	23	.38	38	.63	53	.88
9	.15	24	.40	39	.65	54	.90
10	.17	25	.42	40	.67	55	.92
11	.18	26	.43	41	.68	56	.93
12	.20	27	.45	42	.70	57	.95
13	.22	28	.47	43	.72	58	.97
14	.23	29	.48	44	.73	59	.98
15	.25	30	.50	45	.75	60	1.00

CONVERSION FACTOR: Divide minutes by 60 (minutes in a degree), carrying two places to express quotient as a decimal.

```
Thus,        0.25 decimal of a degree
          60/15.00 minutes
             12 0
             ----
              3 00
              3 00
              ----
                00
```

EXAMPLE: What decimal of a degree is expressed by 15′? Locate 15 in the minute column of table: look to answer in corresponding position in decimal column. Answer: 0.25.

TABLE XI
VARIATION OF HOUSE POWER

House	Weaker Cusp	Stronger Cusp	Variation
6	6.50	7.00	.50
5	7.00	7.50	.50
3	7.50	8.00	.50
2	8.00	8.50	.50
12	8.60	9.30	.70
9	9.30	10.00	.70
8	10.00	10.90	.90
11	10.90	11.90	1.00
4	12.00	14.00	2.00
7	12.50	14.50	2.00
10	13.00	15.00	2.00
House power of M.C. or Asc.:			15.00

EXPLANATION: It has been recognized since ancient times that a planet in an angular house has more power than when it is located in a succedent house or a cadent house, and that a planet in the twelfth house has more power than when it is in the sixth house, even though both are cadent houses. It has also been recognized that in a given house a planet has more power when near the cusp of the house than when farther removed from the house cusp.

Therefore, this table is based not merely on the relative power of each house cusp, but it is also based on the power of an unaspected planet on each cusp relative to the power of the planet derived solely from different aspects.

This table is the basis for calculating the birth-chart power of a planet due solely to its house position. See "Stellar Dynamics."

Table XII

ASTRODYNE VALUE OF THE PARALLEL ASPECT

Min.	Sun, Moon Mercury ANGULAR	Sun, Moon Mercury SUCCEDENT	Sun, Moon Mercury CADENT	M.C., Asc. Planets ANGULAR	Planets SUCCEDENT	Planets CADENT
0	15.00	13.00	11.00	12.00	10.00	8.00
1	14.75	12.78	10.82	11.80	9.83	7.87
2	14.50	12.57	10.63	11.60	9.67	7.73
3	14.25	12.35	10.45	11.40	9.50	7.60
4	14.00	12.13	10.27	11.20	9.33	7.47
5	13.75	11.92	10.08	11.00	9.17	7.33
6	13.50	11.70	9.90	10.80	9.00	7.20
7	13.25	11.48	9.72	10.60	8.83	7.07
8	13.00	11.27	9.53	10.40	8.67	6.93
9	12.75	11.05	9.35	10.20	8.50	6.80
10	12.50	10.83	9.17	10.00	8.33	6.67
11	12.25	10.62	8.98	9.80	8.17	6.53
12	12.00	10.40	8.80	9.60	8.00	6.40
13	11.75	10.18	8.62	9.40	7.83	6.27
14	11.50	9.97	8.43	9.20	7.67	6.13
15	11.25	9.75	8.25	9.00	7.50	6.00
16	11.00	9.53	8.07	8.80	7.33	5.87
17	10.75	9.32	7.88	8.60	7.17	5.73
18	10.50	9.10	7.70	8.40	7.00	5.60
19	10.25	8.88	7.52	8.20	6.83	5.47
20	10.00	8.67	7.33	8.00	6.67	5.33
21	9.75	8.45	7.15	7.80	6.50	5.20
22	9.50	8.23	6.97	7.60	6.33	5.07
23	9.25	8.02	6.78	7.40	6.17	4.93
24	9.00	7.80	6.60	7.20	6.00	4.80
25	8.75	7.58	6.42	7.00	5.83	4.67
26	8.50	7.37	6.23	6.80	5.67	4.53
27	8.25	7.15	6.05	6.60	5.50	4.40
28	8.00	6.93	5.87	6.40	5.33	4.27
29	7.75	6.72	5.68	6.20	5.17	4.13
30	7.50	6.50	5.50	6.00	5.00	4.00
31	7.25	6.28	5.32	5.80	4.83	3.87
32	7.00	6.07	5.13	5.60	4.67	3.73

TABLE XII

ASTRODYNE VALUE OF THE PARALLEL ASPECT

—(Continued)

Min.	Sun, Moon Mercury ANGULAR	Sun, Moon Mercury SUCCEDENT	Sun, Moon Mercury CADENT	M.C., Asc. Planets ANGULAR	Planets SUCCEDENT	Planets CADENT
33	6.75	5.85	4.95	5.40	4.50	3.60
34	6.50	5.63	4.77	5.20	4.33	3.47
35	6.25	5.42	4.58	5.00	4.17	3.33
36	6.00	5.20	4.40	4.80	4.00	3.20
37	5.75	4.98	4.22	4.60	3.83	3.07
38	5.50	4.77	4.03	4.40	3.67	2.93
39	5.25	4.55	3.85	4.20	3.50	2.80
40	5.00	4.33	3.67	4.00	3.33	2.67
41	4.75	4.12	3.48	3.80	3.17	2.53
42	4.50	3.90	3.30	3.60	3.00	2.40
43	4.25	3.68	3.12	3.40	2.83	2.27
44	4.00	3.47	2.93	3.20	2.67	2.13
45	3.75	3.25	2.75	3.00	2.50	2.00
46	3.50	3.03	2.57	2.80	2.33	1.87
47	3.25	2.82	2.38	2.60	2.17	1.73
48	3.00	2.60	2.20	2.40	2.00	1.60
49	2.75	2.38	2.02	2.20	1.83	1.47
50	2.50	2.17	1.83	2.00	1.67	1.33
51	2.25	1.95	1.65	1.80	1.50	1.20
52	2.00	1.73	1.47	1.60	1.33	1.07
53	1.75	1.52	1.28	1.40	1.17	.93
54	1.50	1.30	1.10	1.20	1.00	.80
55	1.25	1.08	.92	1.00	.83	.67
56	1.00	.87	.73	.80	.67	.53
57	.75	.65	.55	.60	.50	.40
58	.50	.43	.37	.40	.33	.27
59	.25	.22	.18	.20	.17	.13

This table is the basis for calculating the birth-chart power of a planet, M.C. or Asc. due solely to the parallel aspect. See "Stellar Dynamics."

TABLE XIII
MUTUAL RECEPTION

Column 1			Column 2			Column 3			Column 4		
	☉		♌		♈		♀		♎	♓	♉
♋		♉		☽		♈	♑	♏		♂	
♊	♒	♍		☿		♐	♋	♓		♃	
♎	♓	♉		♀		♑	♎	♒		♄	
♈	♑	♏		♂		♒		♊		♅	
♐	♋	♓		♃		♓		♐		♆	
♑	♎	♒		♄		♏		♌		♇	
♒		♊		♅			♂		♈	♑	♏
♓		♐		♆		♐	♋	♓		♃	
♏		♌		♇		♑	♎	♒		♄	
	☽		♋		♉	♒		♊		♅	
♊	♒	♍		☿		♓		♐		♆	
♎	♓	♉		♀		♏		♌		♇	
♈	♑	♏		♂			♃		♐	♋	♓
♐	♋	♓		♃		♑	♎	♒		♄	
♑	♎	♒		♄		♒		♊		♅	
♒		♊		♅		♓		♐		♆	
♓		♐		♆		♏		♌		♇	
♏		♌		♇			♄		♑	♎	♒
	☿		♊	♒	♍	♒		♊		♅	
♎	♓	♉		♀		♓		♐		♆	
♈	♑	♏		♂		♏		♌		♇	
♐	♋	♓		♃			♅		♒		♊
♑	♎	♒		♄		♓		♐		♆	
♒		♊		♅		♏		♌		♇	
♓		♐		♆			♆		♓		♐
♏		♌		♇		♏		♌		♇	

INSTRUCTIONS: When the planet in Column 1 is in one of the signs below it, and the planet in column 2 in the space directly to the right of this sign is in one of the signs above it in column 2, the indicated planets are in mutual reception. And when the planet in column 3 is in one of the signs below it, and the planet in column 4 in the space directly to the right of this sign is in one of the signs above it in column 4, the indicated planets are in mutual reception.

EXAMPLE: To determine if a mutual reception forms when the Sun is in Cancer and the Moon in Aries. Find Sun (top column 1), go down to Cancer, now look to planet in column 2 in space directly to the right. Find Moon; check signs at top of column 2. As Aries appears, the two planets are in mutual reception.

TABLE XIV

ESSENTIAL DIGNITY

Planet	Rule Home	Detriment	Exaltation	Deg. Exalt.	Fall	Deg. Fall	Harmony	Inharmony
☉	♌	♒	♈	19	♎	19	♐	♊
☽	♋	♑	♉	3	♏	3	♓	♍
☿	♊-♍	♐-♓	♒	15	♌	15	♏	♉
♀	♎-♉	♈-♏	♓	27	♍	27	♒	♌
♂	♈-♏	♎-♉	♑	28	♋	28	♌	♒
♃	♐-♓	♊-♍	♋	15	♑	15	♉	♏
♄	♑-♒	♋-♌	♎	21	♈	21	♍	♓
♅	♒	♌	♊	7	♐	7	♎	♈
♆	♓	♍	♐	18	♊	18	♋	♑
♇	♏	♉	♌	17	♒	17	♈	♎

Table XV

HARMODYNE AND DISCORDYNE VALUE OF ESSENTIAL DIGNITY

	☉	☽	☿	♀	♂	♃	♄	♅	♆	♇
♈	19°+4 +3			−2	+2		21°−4 −3	−1		+1
♉		3°+4 +3	−1	+2	−2	+1				−2
♊	−1		+2			−2		7°+4 +3	18°−4 −3	
♋		+2			28°−4 −3	15°+4 +3	−2		+1	
♌	+2		15°−4 −3	−1	+1		−2	−2		17°+4 +3
♍		−1	+2	27°−4 −3		−2	+1		−2	
♎	19°−4 −3			+2	−2		21°+4 +3	+1		−1
♏		3°−4 −3	+1	−2	+2	−1				+2
♐	+1		−2			+2		7°−4 −3	18°+4 +3	
♑		−2			28°+4 +3	15°−4 −3	+2		−1	
♒	−2		15°+4 +3	+1	−1		+2	+2		17°−4 −3
♓		+1	−2	27°+4 +3		+2	−1		+2	

This table is the basis for calculating the birth-chart harmodynes or discordynes of a planet due solely to essential dignity. See "Stellar Dynamics."

TABLE XVI

PLANETARY HOURS

The approximate time of sunrise, sunset and length of Planetary Day Hour for each month in latitudes 30°, 40° and 50° North. For Southern Hemisphere, use same time but opposite months to those given.

	April and September			May and August			June and July		
Latitude	30°	40°	50°	30°	40°	50°	30°	40°	50°
Sunrise A.M.	5:41	5:33	5:21	5:22	5:04	4:40	5:02	4:35	3:55
Sunset P.M.	6:19	6:27	6:39	6:38	6:56	7:20	6:58	7:25	8:05
Length hour	1h-3m	1h-4m	1h-6m	1h-6m	1h-10m	1h-13m	1h-10m	1h-14m	1h-21m

	October and March			November and Feb.			December and Jan.		
Latitude	30°	40°	50°	30°	40°	50°	30°	40°	50°
Sunrise A.M.	6:19	6:27	6:39	6:38	6:56	7:20	6:58	7:25	8:05
Sunset P.M.	5:41	5:33	5:21	5:22	5:04	4:40	5:02	4:35	3:55
Length hour	0h-57m	0h-56m	0h-54m	0h-54	0h-51m	0h-47	0h-50	0h-46	0h:39m

For the night hour (between sunset and sunrise) use the day hour of the opposite month—this month being just above or below. Thus for the night hour of April, use day hour given for October.

Planetary rulership of each hour of each day of the week reckoned from sunrise, noon, midnight and sunset.

	Sunrise	2nd	3rd	4th	5th	6th	Noon	2nd	3rd	4th	5th	6th
Sunday	Sun	Venus	Merc.	Moon	Sat.	Jup.	Mars	Sun	Venus	Merc.	Moon	Sat
Monday	Moon	Sat.	Jup.	Mars	Sun.	Venus	Mercury	Moon	Sat.	Jup.	Mars	Sun
Tues.	Mars	Sun	Venus-	Merc.	Moon	Sat.	Jupiter	Mars	Sun	Venus	Merc.	Moon
Wed.	Mercury	Moon	Sat.	Jup.	Mars	Sun	Venus	Merc.	Moon	Sat.	Jup.	Mars
Thurs.	Jupiter	Mars	Sun	Venus	Merc.	Moon	Saturn	Jup.	Mars	Sun	Venus	Merc.
Friday	Venus	Merc.	Moon	Sat.	Jup.	Mars	Sun	Venus	Merc.	Moon	Sat.	Jup.
Sat.	Saturn	Jup.	Mars	Sun	Venus	Merc.	Moon	Sat.	Jup.	Mars	Sun	Venus

	Sunset	2nd	3rd	4th	5th	6th	Midnight	2nd	3rd	4th	5th	6th
Sunday	Jupiter	Mars	Sun	Venus	Merc.	Moon	Venus	Merc.	Moon	Sat.	Jup.	Mars
Monday	Venus	Merc.	Moon	Sat.	Jup.	Mars	Saturn	Jup.	Mars	Sun	Venus	Merc.
Tues.	Saturn	Jup.	Mars	Sun	Venus	Merc.	Sun	Venus	Merc.	Moon	Sat.	Jup.
Wed.	Sun	Venus	Merc.	Moon	Sat.	Jup.	Moon	Sat.	Jup.	Mars	Sun	Venus
Thurs.	Moon	Sat.	Jup.	Mars	Sun	Venus	Mars	Sun	Venus	Merc.	Moon	Sat.
Friday	Mars	Sun	Venus	Merc.	Moon	Sat.	Mercury	Moon	Sat.	Jup.	Mars	Sun
Sat.	Mercury	Moon	Sat.	Jup.	Mars	Sun	Jupiter	Mars	Sun	Venus	Merc.	Moon

INSTRUCTIONS: Find length of planetary hour for month and latitude, and the time of sunrise and sunset. Add duration of hour so found to preceding sunrise, noon, sunset or midnight as many times as is necessary to reach the desired planetary hour or time of day as shown in Table. The first hour from any of the four points lasts until the time found by adding its duration to the time of the point. Second hour may be found by adding the duration of the hour to the end of the first hour, giving the end of the second and beginning of the third hour.

From the above tables the planetary hours of any day of the year may quickly be found with precision enough for all practical purposes.

TABLE XVII

CHARACTER VIBRATIONS

	No.	*Letter*	*Color*	*Tone*	*Gem or metal*
Ari	13	M	Light Red	High C	Amethyst
Tau	14	N	Dark Yellow	Low E	Agate
Gem	17	F, PH, P	Light Violet	High B	Beryl
Can	18	SH, TS, TZ	Light Green	High F	Emerald
Leo	19	Q	Light Orange	High D	Ruby
Vir	2	B	Dark Violet	Low B	Jasper
Lib	3	G	Light Yellow	High E	Diamond
Sco	4	D	Dark Red	Low C	Topaz
Sag	7	Z	Light Purple	High A	Red Garnet
Cap	8	H, CH	Dark Blue	Low G	Onyx
Aqu	9	TH	Sky Blue	High G	Blue Sapphire
Pis	12	L	Dark Purple	Low A	Peridot
Sun	21	S	Orange	D	Gold
Moo	20	R	Green	F	Silver
Mer	1	A	Violet	B	Quicksilver
Ven	6	U, V, W	Yellow	E	Copper
Mar	16	O	Red	C	Iron
Jup	5	E	Purple	A	Tin
Sat	15	X	Blue	G	Lead
Ura	10	I, J, Y	Dazzling White	Astral Chimes	Uranium
Nep	11	C, K	Iridescence	Music of Spheres	Neptunium
Plu	22	T	Ultra-violet, Infra-red	Spirit Choir	Plutonium

The zodiacal signs and planets have been abbreviated with their first three letters to make the table more compact.

VII

Stellar Dynamics

If the principles for each rule were explained and examples given for each calculation used in Stellar Dynamics, a bulky volume would result. "The Astrodyne Manual", by Elbert Benjamine, tells that story.

This chapter includes an introduction to the subject; a brief outline of the mathematical calculation of astrodynes, harmodynes and discordynes; the United States of America Chart and the Aquarian Age Chart, both accompanied by their stellar dynamic grills; and 24 historical events occurring at the time of 24 major progressions in the two charts.

Introduction

"Each soul is attached to the physical form through which it manifests by the psychokinetic energy of the thought-cells of its unconscious mind. These thought-cells compete with the thought-cells of other souls to attach themselves to and govern the development of the seed at the moment of fertilization. After gestation—or in lower life-forms the process which takes the place of gestation—the physical form is born. But to be born, each important group of thought-cells in the unconscious mind must receive additional energy from the planet or sign which has similar vibratory characteristics. These vibratory characteristics include similarity in power and harmony or discord as well as the resonance typical of the planet or sign.

"Receiving energy from such similar vibratory rates gives each group of thought-cells in the finer form that much additional psychokinetic power. And when all important groups of thought-cells gain psychokinetic power through similar astrological energy being added to them, they are able to use that power to cause the life-form at that moment to be born.

"An individual is not born until the inner-plane weather as

mapped by astrological positions is closely similar to the organization of the thought-cells within his unconscious mind. Thus his birth-chart is a very accurate map of his character at the time of his birth. The power of each planet indicates the power of the thought-cells it maps, and each aspect received by a planet indicates whether the thought-elements ruled by the other planet involved in the aspect have entered into combination with the thought-elements ruled by this planet harmoniously, discordantly or neutrally.

"The relative power of a planet indicates the relative power of the thought-cells of that planetary type in the unconscious mind, and thus the relative power of the natural aptitudes characteristic of the planet. The more powerful the natural aptitudes the easier it is to develop them into ability of their particular planetary type. But in addition to influencing the thoughts and behavior, the thought-cells possess psychokinetic power by which they bring into the life conditions and events of the kind they desire. Therefore, the more powerful a planet is in the birth-chart, the more power the thought-cells it maps have to bring into the life conditions and events of its type.

"The more harmony a planet has, the more fortunate are the thoughts and behavior the thought-cells it maps influence, and the more fortunate are the events which the thought-cells it maps use their psychokinetic power to bring into the life." More discord a planet has, the more unfortunate are the events attracted.

"The actual events which thought-cells can bring into the life depend upon two things: the psychokinetic power of the thought-cells, and the resistance offered by the physical environment to the kind of events the thought-cells try to bring to pass. An environment of the same planetary type as the thought-cells not only facilitates bringing the event they desire to pass through offering little resistance, but it also, by its inner-plane radiations, increases the psychokinetic power of such thought-cells. Thus while the relative power of the thought-cells as mapped by the power of the planets in the birth-chart, indicates the natural abilities an individual can most easily develop and use, the relative harmony of the planets in the birth-chart indicates in association with what kind of environment he can use these abilities with most fortunate results.

"According to their relative power and harmony the planets not only show the abilities and environment in which they can most successfully be used, but they also indicate the events and diseases of a particular type toward which there is a predisposition. Therefore it is very important to know as precisely as possible both the power of each planet in the birth-chart and its harmony or discord.

"The planets map the dynamic thought-cell structures, and the signs ruled by a planet map the common thought-cells ruled by the planet. The houses map the compartments of the inner-plane form, and the signs on their cusps, or intercepted in them, together with any planet in them, indicate their power and harmony or discord. To the extent a sign is powerful does association with the things it rules become important in the life. And to the extent a sign is harmonious or discordant does association with the things it rules attract fortune or misfortune, and is there a predisposition for the department of life it governs to be fortunate or unfortunate.

"Therefore, to know which environmental conditions to seek and which to shun, and to know into which departments of life to channel the energies, and which to avoid, and relative to which to use special precautionary measures, it is essential the individual should know, as precisely as possible, both the power of each sign and house of the birth-chart, and its harmony or discord.

"However, the birth-chart alone does not determine the conditions and events that enter an individual's life. The outer-plane environment and the inner-plane environment also have an influence over them. His thoughts and behavior react in a certain way both to the outer-plane and to the inner-plane environment, according to his thought-cell organization as mapped at birth and modified by later experience. The inner-plane environment has as much influence over the individual as does the outer-plane environment, and the chief influence of the inner-plane environment is that of the inner-plane weather mapped by progressed aspects.

"The more precisely the individual knows the manner in which his thought-cells will react to any given type of inner-plane weather, and as a result use their psychokinetic power to bring events and conditions of a particular kind into his life, the better he is able to know just what precautionary actions are necessary to overcome a given period of adverse inner-plane weather or to take advantage of a given period of favorable inner-plane weather. This is shown by the power and harmony of the planets in his chart of birth."

In spite of any mathematical system to guage it, "the birth-chart only maps the power and harmony and discord of the thought-cells at birth, and how they are apt to react to inner-plane weather of a particular kind. It only indicates the general precautionary actions that should be taken. But in addition, if the individual is to get as much benefit from life as he should, he must know the precise kind of inner-plane weather he will experience during each period of his life. And to gauge the amount and kind of precautionary actions necessary, he should not only know in advance the type of inner-plane weather, as indicated by the planets involved in the progressed aspect, and whether it will be favorable or unfavorable, as indicated by the nature of the aspect, but he should also know its intensity, and how favorable or unfavorable it will be."

The foregoing quote by Elbert Benjamine embraces a concise summary of the principal research findings regarding the birth-chart. When these factors became clear, The Church of Light Astrological Research Department began seriously to study timed birth-charts in view of finding formulas to determine the relative amounts of power, harmony and discord in a birth-chart.

The solutions to four problems were sought. Find a formula to be used with any given chart which will accurately express in mathematical terms:

I—The amount of energy mapped by each brith-chart planet, the Midheaven and Ascendant, and which can be used to accurately express the amount of energy mapped by any progressed aspect.

II—The amount of energy mapped by each brith-chart house and each zodiacal sign.

III—The amount of harmony and/or discord mapped by each birth-chart planet, Midheaven and Ascendant, and which can be used to accurately express the amount of harmony and/or discord mapped by any progressed aspect.

IV—The amount of harmony and/or discord mapped by each birth-chart house and each zodiacal sign.

The problem of probability mathematics is one the critics of astrology must solve before any of their criticism can be taken seriously. So far, we have been unable to solve this problem in the rigorous manner to be desired, and perhaps it never will be solved in a strict sense due to the very nature of astrology, which—like the social sciences—deals with individual differences and individuals within groups.

In struggling to find the formulas, many difficulties were encountered. Over the years several mathematical systems were employed, then discarded; each one evolving into another until the development of the present method of Stellar Dynamics, a mathematical method employed to determine the number of astrodynes (units of astrological power), harmodynes (units of astrological harmony), and discordynes (units of astrological discord) represented by each birth-chart aspect, planet, sign and house, and each progressed aspect. This method indicates precisely the relative power, harmony and discord of the thought-cells mapped by each planet, sign and house, but IT DOES NOT REPRESENT AN ABSOLUTE MEASUREMENT.

Another important factor is that evolutionary level, conditioning and personal effort at improving character modifies what a cold column of figures might otherwise indicate. The terms used in the field of relativity adequately explain why stellar dynamics does not give an absolute measurement. Astrodynes, harmodynes and discordynes work on the evolutionary level, but at the same time they cannot be compared with those in some other chart, even on the same evolutionary level. Each chart operates within its own frame of reference, and the frames of reference of any two charts are different. Therefore the total number of astrodynes (power) alone in a chart does not indicate whether a person can or cannot develop great ability. A study of the astrodynes in the charts of the U. S. Presidents interprets in a mathematical manner the meaning of that statement.

George Washington	477.31	Andrew Johnson	544.43
John Adams	588.71	Ulysses S. Grant	439.37
Thomas Jefferson	393.21	Rutherford B. Hayes	452.86
James Madison	555.07	James A. Garfield	533.04
James Monroe	484.16	Chester A. Arthur	474.44
John Quincy Adams	567.39	Grover Cleveland	447.86
Andrew Jackson	404.62	Benjamin Harrison	572.95
Martin Van Buren	669.18	William McKinley	628.14
William Henry Harrison	719.81	Theodore Roosevelt	471.38
John Tyler	564.04	William Howard Taft	684.26
James K. Polk	468.94	Woodrow Wilson	638.81
Zachary Taylor	609.49	Warren G. Harding	617.46
Millard Fillmore	545.75	Calvin Coolidge	477.92
Franklin Pierce	537.30	Herbert C. Hoover	736.30
James A. Buchanan	500.94	Franklin D. Roosevelt	488.47
Abraham Lincoln	485.11	Harry S. Truman	452.65

William Henry Harrison, the nation's ninth president, died after one month in office; his chart shows 719.81 astrodynes, a larger total than any other excepting that of Herbert C. Hoover. Yet Thomas Jefferson, who left a great impress upon our culture, had a chart showing only 393.21 astrodynes. A study of these two charts in conjunction with their case histories showing conditioning energy demonstrates rather spectacularly why stellar dynamics is not an absolute measurement. Two more factors are indicated: (1) the impossibility of arriving at a mathematical method giving an absolute measurement to cover any and all charts; and (2) the great value of stellar dynamics when associated with one frame of reference.

Basic Principles

Precision with which natal or progressed stellar dynamics may be ascertained depends upon precision used in erecting the birth-chart. It is only when a timed birth-chart is calculated to show the positions of the ten planets, Midheaven and Ascendant—both zodiacally and by declination—and all minor cusps calculated to the nearest minute that natal or progressed power and harmony in a chart can be ascertained with equal precision. Therefore, an accurately timed and erected birth-chart is the first requisite.

Calculation of astrodynes (power) is based upon three things:

1. *Unmodified power of the planet.* Research indicates Sun, Moon and Mercury have more power when occupying a given natal position than do any of the other seven planets; thus the unmodified power of Sun, Moon and Mercury is greater.

2. *Birth-chart house occupied by the planet.* A recognized astrological principle is that a planet stationed in an angular house has more power than the same planet in a succedent or cadent house; thus aside from its unmodified power and aspects, the power of a planet varies according to the birth-chart house it occupies.

3. *Natal aspects of the planet.* A generally accepted astrological principle is that each aspect a planet forms increases its power; the amount of the power gained varying according to aspect. (Table VII: ASPECTS AND THEIR ORBS)

All subsequent calculations of astrodynes are based upon the number of astrodynes possessed by the natal planets, representing the most active of the four basic astrological factors: planets, signs, aspects and houses.

Calculation of harmodynes (harmony) and discordynes (discord) is based upon three things:

1. *Sign the planet occupies.* Values given for essential dignity: Mutual Reception, 5 harmodynes; Degree of Exaltation, 4 harmodynes; Sign of Exaltation, 3 harmodynes; Home Sign, 2 harmodynes; Sign of Harmony, 1 harmodyne; Sign of Inharmony, 1 discordyne; Sign of Detriment, 2 discordynes; Sign of Fall, 3 discordynes; Degree of Fall, 4 discordynes. (See Table of Mutual Reception; Table of Essential Dignity; Table of Harmodyne and Discordyne Value of Essential Dignity.)

2. *Nature of planet forming the aspect.* Values given when

a planet is involved in any one of the ten aspects: Jupiter, ½ power of aspect is harmonious; Venus, ¼ power of aspect is harmonious; Mars, ¼ power of aspect is discordant; Saturn, ½ power of aspect is discordant. Remaining planets are neutral in this respect.

3. *Nature of aspect between two planets.* Three types: (a) power of *harmonious* aspects—trine, sextile, semi-sextile—expressed as astrodynes equals same number of harmodynes; (b) power of *discordant* aspects—opposition, square, sesqui-square, semi-square—expressed as astrodynes equals same number of discordynes; (c) power of *neutral* aspects—conjunction, inconjunct, parallel—expressed as astrodynes add or subtract nothing as far as harmony and discord are concerned.

Outline of Stellar Dynamics

1. *HOW TO CALCULATE THE POWER GRILL*

(1) Aspects except parallel: (For Mercury, see D.)
- A Distance from perfect aspect.
- B Power of perfect aspect. Always use larger orb. (See Orb Table.)
- C Subtract A from B, transposing minutes into decimal fractions of a degree. (See Table Decimals Of A Degree.)
- D Mercury aspects:
  - a Distance from perfect aspect. (Determined by *planetary* orb.)
  - b Power of perfect aspect. (Determined by *luminary* orb for house Mercury occupies.)
  - c Subtract a from b, transposing minutes as above.

(2) Parallel Aspects:
- A Minutes from perfect aspect.
- B Look up A in Table Astrodyne Value Of Parallels for answer.

(3) House Position:
- A Planet's distance from nearest cusp; convert to minutes.
- B Number of degrees in house; convert to minutes.

C Divide A by B, carry to two decimal places.

D Variation for house power. (See Table Variation Of House Power.)

E Multiply C by D.

F Add E to power of weaker cusp, or subtract from power of stronger cusp, whichever was used in calculating A. (See Table Variation Of House Power for values of house cusps.)

2. *HOW TO CALCULATE THE HARMONY/DISCORD GRILL*

(1) Aspects:

A Power of Aspect.

B Modify for nature of aspect. (Plus sign indicates harmodynes; minus sign, discordynes.)

C Modify for nature of planets involved in aspect.

(2) Mutual Reception. (See Table Of Mutual Reception.)

(3) Essential Dignity. (See Table Harmodyne And Discordyne Value of Essential Dignity.)

(4) Total all harmodynes; then all discordynes.

(5) To get net harmony or discord, substract the smaller from the larger figure, taking the sign of the larger.

3. *HOW TO CALCULATE POWER OF THE SIGNS*

(1) Rulership (unoccupied) power of intercepted sign is ¼ power of ruler (s).

(2) Rulership (unoccupied) power is ½ power of its ruler (s) for each house cusp occupied.

RULE: Add unoccupied power to the total power of each planet in the sign. (Include M.C. and Asc.)

4. *HOW TO CALCULATE HARMONY/DISCORD OF THE SIGNS*

(1) Rulership (unoccupied) harmony/discord of intercepted sign is ¼ net harmony/discord of ruler(s).

(2) Rulership (unoccupied) harmony/discord is ½ net harmony/discord for each house cusp occupied.

RULE: Find net of rulership harmony/discord and total harmony/discord of each planet in the sign.

5. *HOW TO CALCULATE POWER OF THE HOUSES*

(1) Rulership (unoccupied) power of sign on cusp, ½ of ruler(s).
(2) Rulership (unoccupied) power of intercepted sign(s), if any, ¼ of ruler(s).
(3) Total power of all planets in house.
(4) In tenth and first house, include M.C. and Asc., respectively.

RULE: Add (1), (2), (3), and (4).

6. *HOW TO CALCULATE HARMONY/DISCORD OF THE HOUSES*

(1) Rulership harmony/discord is ½ H/D of ruler of sign on cusp.
(2) Rulership H/D is ¼ ruler(s) of intercepted sign, if any.
(3) Net H/D of all planets in the house.
(4) In 10th and 1st houses, include M.C. and Asc., respectively.

RULE: Add all H; add all D; subtract the smaller from the larger, taking the sign of larger, to get net H/D.

The above outline—not meant to teach Stellar Dynamics—was designed as a reminder to one who understands fundamentals of Stellar Dynamics, much as a map guides a person on the right road. It covers only the subject of the birth-chart.

Major progressions, minor progressions and transit progressions can also be calculated in astrodynes, harmodynes and discordynes. Effective use is found when the peak power of an aspect, planet, house or sign is needed when starting some venture of importance. The rules are explained in detail in "The Astrodyne Manual", by Elbert Benjamine, and in "Stellar Healing", by C. C. Zain.

The system of Stellar Dynamics can be applied to all charts for precision. Better results were found when it was employed than merely guessing at the factors of strength, harmony and discord in a chart.

669 United States of America

July	4	1776
(Month	(Day)	(Year)

(Name)

Place Philadelphia, Pa.

Latitude 39 N 57

Longitude 75 W 08

DOMINANT FACTOR

Time of Birth (Daylight Saving) ____

Correction for Standard Time ____

Time of Birth (Standard Time) ____

Correction for Mean Time ____

Local Mean Time of Birth, A.M or P.M. 02:14:43

FIRST KEY PROBLEM

Noon 11:59:60 12:00

Local Mean Time 02:14:43

L.M.T. Interval 09:45:17 Minus

Sidereal Time 06:51:48

(Noon) 24:00:00

30:51:48

L.M.T. Interval 09:45:17 Minus

S. T. (Uncorrected) 21:06:31

Correction, 9.86s per h. for E.G.M.T. Int. 00:47

Sidereal Time (Of Birth) 21:05:44

SECOND KEY PROBLEM

Local Mean Time of Birth 2:14:43 A.M.

Hrs. and mins. E. or W of Greenwich 5:00:32 W

E.G.M.T. 7:15:15

Noon 11:59:60 12:00

E.G.M.T. 7:15:15

E.G.M.T. Interval (Indicate plus or minus) 4:44:45 Minus

Parker's Ephemeris

Dalton's Table of Houses

ADDITIONAL FACTORS

Constant Log .7035

Limiting Date (Including year) Sept. 14, 1776

Doris Chase Doane

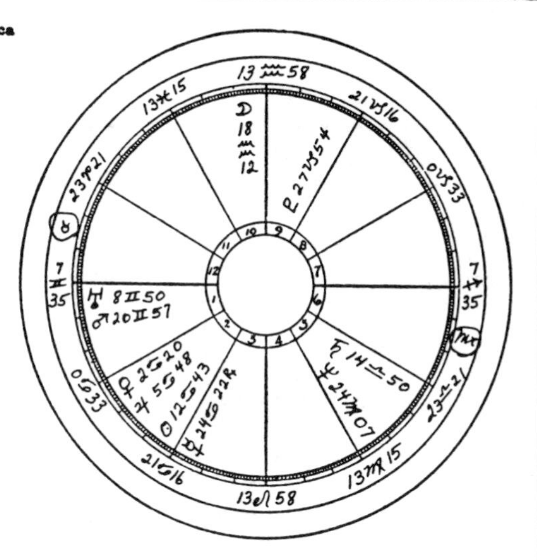

Mov.	Fix.	Mut.	Fire	Earth	Air	Water	Ang.	Suc.	Cad.
6	M.C. 1	Asc. 3	0	2	M.C 4 Asc	4	3	5	2
Per.	**Comp.**	**Pub.**	**Life**	**Wealth**	**Assoc.**	**Psy.**	**Above**	**East**	**Ret.**
6	2	2	5	4	1	0	2	7	1

Declinations	ASPECTS													
		☉	☽	☿	♀	♂	♃	♄	♅	♆	♇	MC	ASC	
22 N 51	☉		.	☌	☌ P	P	☌ P	□	.	.	.	⚻	.	
16 S 16	☽			.	Q	△	Q	△	△	.	.	☌ P	△	
17 N 35	☿				.	.	.	.	∠	⚹	☍	P	∠	
23 N 29	♀					☌ P	☌ P	.	.	.	P	Q	.	
23 N 31	♂						P	△	.	□	P	△	.	
23 N 16	♃							.	.	.	P	.	⚺	
3 S 31	♄								△	P	.	△	△	
21 N 45	♅									.	Q	△	☌ P	
4 N 07	♆										△	Q	.	
24 S 00	♇											.	△	
16 S 37	M.C.												△	
21 N 37	Asc.													

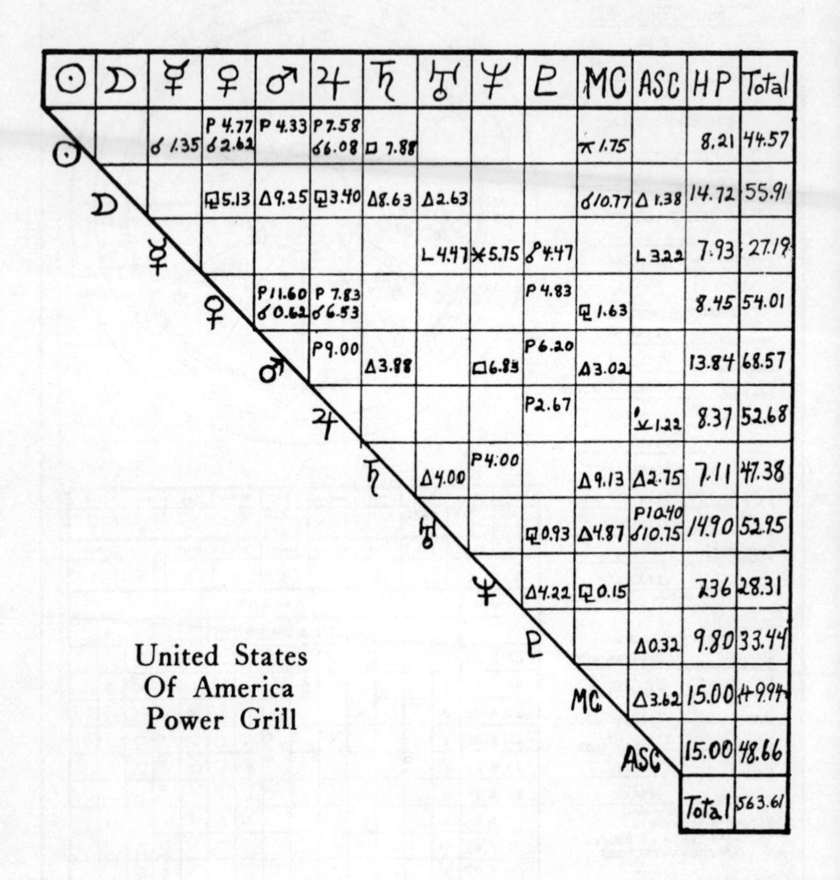

	☉	☽	☿	♀	♂	♃	♄	♅	♆	♇	MC	ASC	HP	Total
☉			☌ 1.35	P 4.77 ☌ 2.62	P 4.33	P 7.58 ☌ 6.08	□ 7.88				⚻ 1.75		8.21	44.57
☽				⚼ 5.13	△ 9.25	⚼ 3.40	△ 8.63	△ 2.63			☌ 10.77	△ 1.38	14.72	55.91
☿								∠ 4.47	⚹ 5.75	☍ 4.47		∠ 3.22	7.93	27.19
♀					P 11.60 ☌ 0.62	P 7.83 ☌ 6.53				P 4.83	⚼ 1.63		8.45	54.01
♂						P 9.00	△ 3.88		□ 6.83	P 6.20	△ 3.02		13.84	68.57
♃										P 2.67		⚺ 1.22	8.37	52.68
♄								△ 4.00	P 4.00		△ 9.13	△ 2.75	7.11	47.38
♅										⚼ 0.93	△ 4.87	P 10.40 ☌ 10.75	14.90	52.95
♆										△ 4.22	⚼ 0.15		7.36	28.31
♇												△ 0.32	9.80	33.44
MC												△ 3.62	15.00	49.94
ASC													15.00	48.66
													Total	563.61

United States
Of America
Power Grill

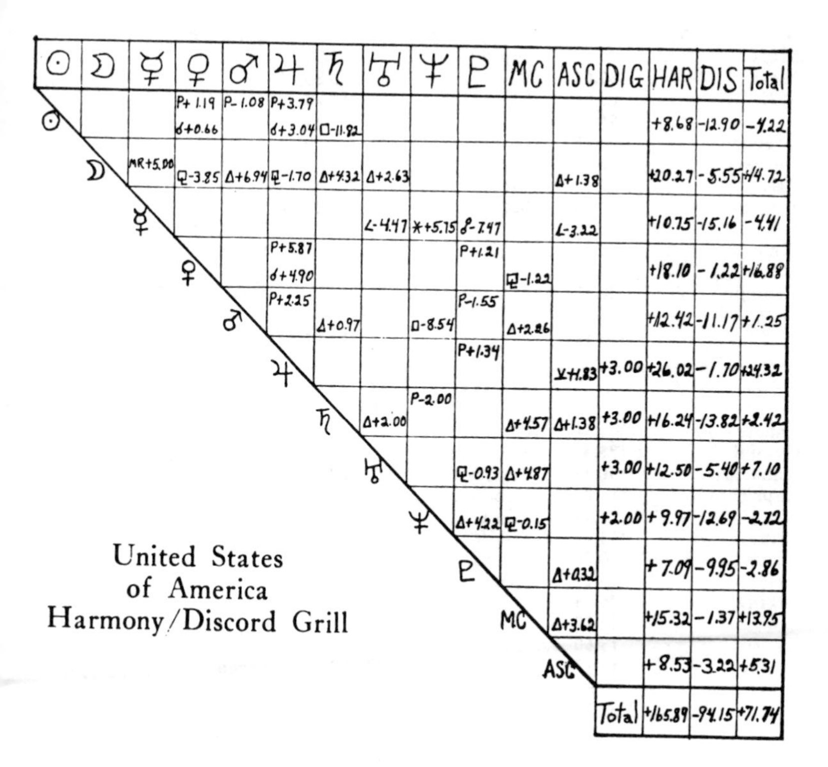

	☉	☽	☿	♀	♂	♃	♄	♅	♆	♇	MC	ASC	DIG	HAR	DIS	Total
☉				P+1.19 ☌+0.66	P-1.08	P+3.79 ☌+3.04	□-11.82							+8.68	-12.90	-4.22
☽			MR+5.00	⚼-3.85	△+6.94	⚼-1.70	△+4.32	△+2.63				△+1.38		+20.27	-5.55	+14.72
☿								∠-4.47	✱+5.15	☍-7.47		∠-3.22		+10.75	-15.16	-4.41
♀						P+5.87 ☌+4.90				P+1.21	⚼-1.22			+18.10	-1.22	+16.88
♂						P+2.25	△+0.97		□-8.54	P-1.55	△+2.26			+12.42	-11.17	+1.25
♃										P+1.34		⚺+1.83	+3.00	+26.02	-1.70	+24.32
♄								△+2.00	P-2.00		△+4.57	△+1.38	+3.00	+16.24	-13.82	+2.42
♅										⚼-0.93	△+4.87		+3.00	+12.50	-5.40	+7.10
♆										△+4.22	⚼-0.15		+2.00	+9.97	-12.69	-2.72
♇												△+0.32		+7.09	-9.95	-2.86
MC												△+3.62		+15.32	-1.37	+13.95
ASC														+8.53	-3.22	+5.31
													Total	+165.89	-94.15	+71.74

United States
of America
Harmony/Discord Grill

Stellar Dynamics — U. S. A. Chart

THE PLANETS, MIDHEAVEN AND ASCENDANT

ASTRODYNES		*HARMODYNES*		*DISCORDYNES*	
Mars	68.57	Jupiter	24.32		
Moon	65.66	Venus	16.88		
M.C.	60.09	Moon	14.72		
Venus	54.01	M.C.	13.95		
Uranus	52.95	Uranus	7.10		
Jupiter	52.68	Asc.	5.31		
Asc.	48.66	Saturn	2.42		
Saturn	47.38	Mars	1.25		
Sun	45.57			Neptune	2.72
Pluto	36.44			Pluto	2.86
Mercury	30.59			Sun	4.22
Neptune	28.31			Mercury	4.41
	590.91		85.95		14.21
				Net Harmony	71.74

THE SIGNS

Cancer	248.51	Cancer	47.29		
Gemini	185.48	Aquarius	31.05		
Aquarius	150.84	Sagittarius	12.16		
Capricorn	83.82	Gemini	11.45		
Libra	74.39	Libra	10.86		
Virgo	43.60	Pisces	5.40		
Pisces	40.50	Taurus	4.22		
Aries	34.29	Aries	0.63		
Sagittarius	26.34			Scorpio	0.20
Leo	22.79			Capricorn	0.44
Taurus	13.50			Leo	2.11
Scorpio	13.13			Virgo	4.92
	937.19		123.06		7.67
				Net Harmony	115.39

THE HOUSES

One	185.48	Two	44.34		
Two	185.09	Ten	31.05		
Ten	150.84	Seven	12.16		
Five	90.98	One	11.45		
Nine	60.13	Six	8.24		
Three	63.41	Eleven	5.40		
Twelve	47.79	Twelve	4.85		
Eleven	40.50	Three	2.95		
Six	40.14	Eight	1.21		
Seven	26.34			Nine	1.65
Eight	23.69			Four	2.11
Four	22.79			Five	2.50
	937.19		121.65		6.26
				Net Harmony	115.39

United States of America Chart

Source: Ascendant rectified by H. V. Herndon.

Date: Elbert Benjamine started researching this chart in 1908 and used it successfully in his popular mundane predictions appearing in magazines for over a quarter of a century. In this volume, the chart is precisely erected for the Ascendant, Longitude and Latitude given. More than 8 data are used to calculate The USA chart by various astrologers, but paintaking research for almost a half century forces the Research Department to accept the following:

UNITED STATES OF AMERICA

b. July 4, 1776, Philadelphia, Pa., 39N57 75W08, 02:14:43. LMT

Historical Events and Major Progressions

Events: Not selected for any particular reason; picked at random to demonstrate progressions.

1807—Aaron Burr, intending settlement of Western lands or invasion of Mexico, leads conspiracy and is tried for treason: Mars, ruler of 12th (conspiracy), conjunction Sun r, ruler of 4th (lands).

1812, June 18—War of 1812, ostensibly over violation of rights of neutral trade and impressment, but also with desire of repressing Indians through acquisition of Canada: Jupiter, ruler of 7th (war), conjunction Sun r, ruler of 4th (lands); Venus conjunction Sun r, ruler of 4th (lands); Sun semi-square Venus r, ruler of 6th (laborers); Mars (strife), square Saturn r, ruler of 9th (shipping).

1823, Dec. 2—Monroe Doctrine declares America not subject to settlement by European powers: Sun, ruler of 4th (lands), in 4th (lands), semi-square Sun r.

1832—Oregon trail for settlement of Oregon: Mercury in 4th (lands), semi-square Mercury r in 3rd (trail or road).

Black Hawk War: Sun in 4th (lands of Indians and wanted by Whites), sextile Jupiter r in 7th (war).

1846, May 13—War with Mexico, resulting in acquiring western territory; Texas annexed previous year: Sun, ruler of 4th (territory), square Mars r (strife): Sun, ruler of 4th (territory), sextile Jupiter p, ruler of the 7th (war).

1848, Jan. 24—Gold discovered in California: Jupiter in 2nd (cash), sextile Mars r in 1st (people), square Saturn p (Planet of mining). This starts a vast throng traveling west: Mercury, ruler of 3rd (travel), sextile Mars p in 3rd (travel).

1849-50-51—Boom days in western gold fields: Sun conjunction Neptune r (booms), in 5th (speculation).

1861, Apr. 19—First blood of Civil War: Sun square Jupiter r, ruler of 7th (war); Mercury, ruler of 1st (people), in 6th (slaves), sesqui-square Uranus p, in 1st (people).

1871, Apr. 20—Klu-Klux Act is passed giving civil and political rights to negroes: Sun conjunction Saturn r, ruler of 9th (laws); Venus, ruler of 6th (inferiors), trine Venus r, ruler of 12th (secret societies).

1873, Sept. 18—Panic starts, causing several years depression: Venus trine Venus r in 2nd (money); M.C. (business) square Moon r, in 10th (business).

1878-1879—Worst epidemic of Yellow Fever in U. S. history: Venus p in 6th (illness) inconjunct Uranus in 1st (people); Saturn p (restrictions) ruler of 8th (death) conjunction Neptune r.

1893, June 27—Commercial panic is started: Sun trine Jupiter r in 2nd (money); Sun sesqui-square Venus r in 2nd (money); M.C. (business) conjunction Uranus r (disruption).

1894—Pullman Co. strike added to industrial chaos, spreading over 27 states and territories from Chicago and Cincinnati to San Francisco: Sun p in 6th (labor) inconjunct Uranus r (strikes) in 1st (people); Saturn p in 6th (labor) square Mercury r in 3rd (transportation).

1898, Apr. 25—War with Spain: Sun inconjunct Uranus p in 1st (people); Sun semi-square Neptune p; Mars (strife) square Asc. r, (people); Jupiter, ruler of 7th (war), conjunction Mercury r, ruler of 1st (people).

1907, Dec. 16—U. S. Navy sails on world tour with a fleet of battleships: Neptune p (sea) trine Pluto r in 9th (long journeys); Saturn p in 5th (hazards) square Pluto r in 9th (long journeys); Venus p, ruler of 6th (armed services) sextile Saturn r; Mercury p, ruler of 3rd (travel) trine Venus r (armed services).

1910, July 21, Severe forest fires near Canadian border: Saturn p (losses on land) square Pluto r; Moon p (people) in 8th (death) opposition (separation) Sun (homes) in 2nd (money).

1913, Jan. 1—Parcels Post system inaugurated throughout the U. S.; Jupiter p conjunction Mercury r (mail) in 3rd (mail); Mercury (mail) inconjunct Asc. r (people).

1913-17—Many violent strikes. Labor demands 8-hour day and minimum wage. 1916 critical year: Venus p, ruler of 2nd and 6th (money and labor) square Neptune r and inconjunct Mercury (people, travel and hazards); Venus then passed on to make the semi-sextile Pluto r (groups), co-ruler of the 6th (labor).

1917, Apr. 6—U. S. enters World War: Mercury inconjunct Uranus p, ruler of 1st (people); Mars (strife) inconjunct Moon r in 10th (government); Jupiter, ruler of 7th (war) conjunction Mercury r, ruler of 1st (people).

1920, Aug. 26—Woman suffrage went into effect: Moon p (women), ruler of the 10th (government) opposition Sun p; Venus p (women) opposition Venus r.

1924, Nov. 30—Transatlantic photos sent between New York and London: Mercury p (messages) inconjunct Mars r; Jupiter p sextile Neptune r (photos); Neptune p (images) trine Pluto r (electronics).

1925, March 18—Tornado in Missouri and southern Illinois kills 1,000: Sun p (homes) opposition drastic Uranus r (people); Mercury p (people and winds) sextile Mars p (strife); Venus in 8th (death) inconjunct Ascendant (people).

1929, Sept. 19—First great crash in the stock market: Venus p in 8th (other people's money), ruler of 2nd (people's money) opposition (separation) Sun r in 2nd (money); Mars p conjunct Neptune r in 5th (stock market); Jupiter p, ruler of 2nd (money) conjunct Mercury r and sextile Neptune r in 5th (speculation).

1933, Apr. 12—Two billion dollar farm mortgage relief and refinance bill passes house: Roosevelt's congressional message asks 2 billion dollars for home owners' mortgage relief: Moon p, ruler of 2nd (money) trine Sun r, ruler of 4th (homes and farms).

670 Aquarian Age
(Name)

Jan. 19 1881
(Month) (Day) (Yr.)

Place Washington, D.C.

Latitude 38 N 53

Longitude 77 W 01

DOMINANT FACTOR

Time of Birth (Daylight Saving)

Correction for Standard Time

Time of Birth (Standard Time)

Correction for Mean Time

Local Mean Time of Birth, A.M or P.M 15:48:24

FIRST KEY PROBLEM

Noon 12:00

Local Mean Time

L.M.T. Interval 03:48:24 Plus

Sidereal Time (Noon) 19:56:04

L.M.T. Interval 03:48:24

S. T. (Uncorrected) 23:44:28

Correction, 9.86s per h. for E.G.M.T. Int. 01:28

Sidereal Time (Of Birth) 23:45:56

SECOND KEY PROBLEM

Local Mean Time of Birth 3:48:24 P.M.

Hrs. and mins. E. or W. of Greenwich 5:08:04 W

E.G.M.T. 8:56:28 P.M.

Noon 12:00

E.G.M.T.

E.G.M.T. Interval 8:56 Plus (Indicate plus or minus)

Die Deutsche Ephemerides

Dalton's Tables of Houses

ADDITIONAL FACTORS

Constant Log .4292

Limiting Date Sept. 5. 1880 (Including year)

Kathleen Robinson

Chart wheel: ♈ 26♓10; 27♒44; 4♒58; 14♑52; 10♐47; 2♏05; ♎ 26♍10; 27♌44; 4♌58; 14♋52; 10♊47; 2♉05.
♅ ♄ 13♈07, 22♈45; ♆ ♇ 11♉31, 26♉24 R; ♀ 14♓06; ☉ 0♒00; ☿ 25♑45; ♂ 4♑04; ♅ ☽ 13♍20 R, 18♍38.

Mov.	Fix.	Mut.	Fire	Earth	Air	Water	A(ng).	Suc.	Cad.
Asc 4	3	M.C 3	2	6	1	M.C 1 Asc	4	2	4
Per.	**Comp.**	**Pub.**	**Life**	**Wealth**	**Assoc.**	**Psy.**	**Above**	**East**	**Ret.**
2	3	5	1	3	6	0	7	6	2

ASPECTS

Declinations		☉	☽	☿	♀	♂	♃	♄	♅	♆	♇	MC	ASC
20 S 10	☉		⚼	☌	∠	.	.	□	⚼	□	△	⚹	'
0 S 14	☽			△	☍	.	.	.	☌	△	△	☍	⚹
22 S 52	☿				∠	.	.	□	⚼	.	△	⚹	☍ P
7 S 06	♀					.	⚺	P	☍ P	⚹	P	.	△
23 S 58	♂						□	.	.	△	∴	□	☍
4 N 03	♃							☌	⚻	⚺	∠	.	□
6 N 34	♄								P	.	P	.	□
7 N 18	♅									△	.	.	⚹
13 N 36	♆										.	∠	⚹
6 N 06	♇											⚹	∠
1 S 32	M.C.												.
22 N 38	Asc.												

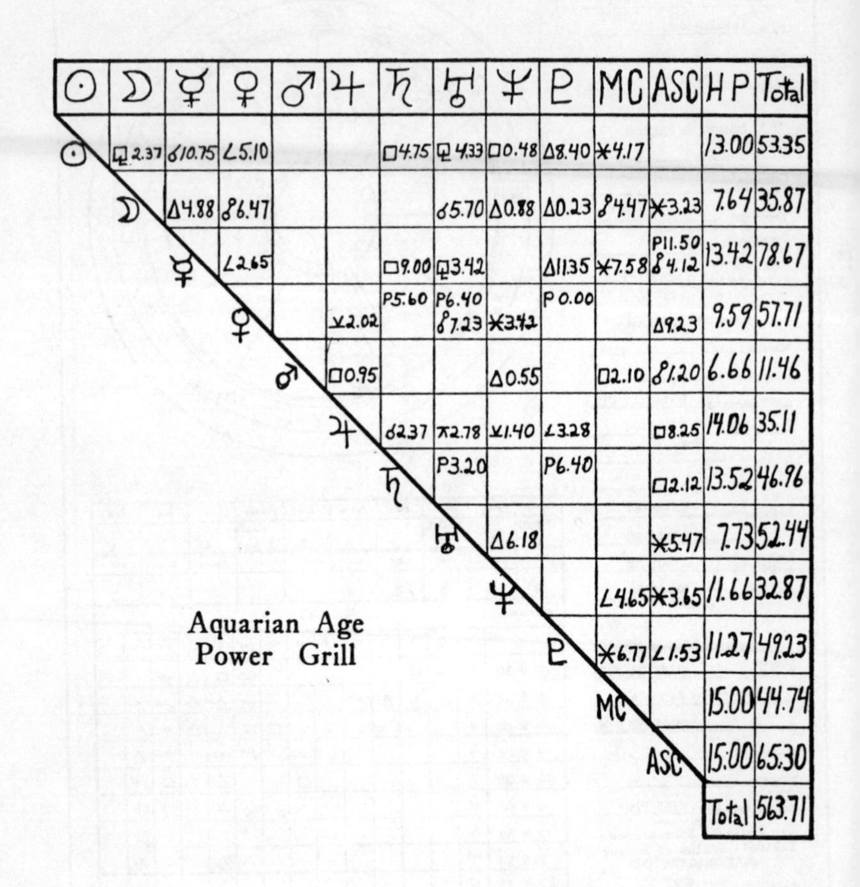

	☉	☽	☿	♀	♂	♃	♄	♅	♆	♇	MC	ASC	HP	Total
☉		Q2.37	☌10.75	∠5.10			□4.75	Q4.33	□0.48	△8.40	✱4.17		13.00	53.35
☽			△4.88	☍6.47				☌5.70	△0.88	△0.23	☍4.47	✱3.23	7.64	35.87
☿				∠2.65			□9.00	Q3.42		△11.35	✱7.58	P11.50 ☌4.12	13.42	78.67
♀						⊻2.02	P5.60	P6.40 ☍7.23	✱3.42	P0.00		△9.23	9.59	57.71
♂						□0.95			△0.55		□2.10	☍1.20	6.66	11.46
♃							☌2.37	⊼2.78	⊻1.40	∠3.28		□8.25	14.06	35.11
♄								P3.20		P6.40		□2.12	13.52	46.96
♅									△6.18			✱5.47	7.73	52.44
♆											∠4.65	✱3.65	11.66	32.87
♇											✱6.77	∠1.53	11.27	49.23
MC													15.00	44.74
ASC													15.00	65.30
													Total	563.71

Aquarian Age
Power Grill

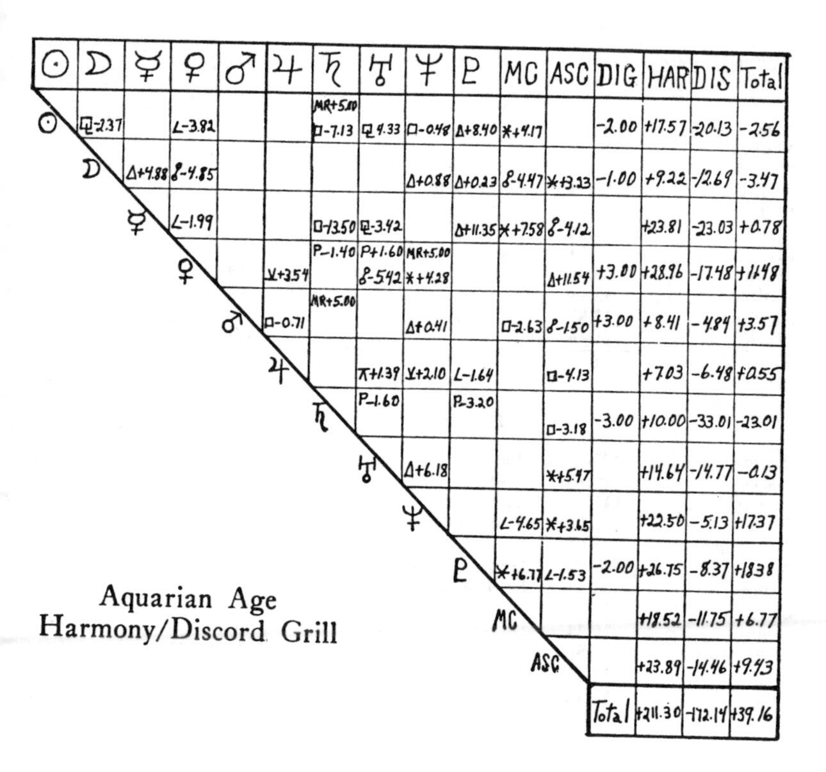

	☉	☽	☿	♀	♂	♃	♄	♅	♆	♇	MC	ASC	DIG	HAR	DIS	Total
☉		⚼-2.37		∠-3.82			MR+5.00 □-7.13	⚼4.33	□-0.48	△+8.40	✶+4.17		-2.00	+17.57	-20.13	-2.56
☽			△+4.88	☍-4.85					△+0.88	△+0.23	☍-4.47	✶+3.23	-1.00	+9.22	-12.69	-3.47
☿				∠-1.99			□-13.50	⚼-3.42		△+11.35	✶+7.58	☍-4.12		+23.81	-23.03	+0.78
♀						⚺+3.54	P-1.40	P+1.60 ☍-5.42	MR+5.00 ✶+4.28			△+11.54	+3.00	+28.96	-17.48	+11.48
♂						□-0.71	MR+5.00		△+0.41		□-2.63	☍-1.50	+3.00	+8.41	-4.84	+3.57
♃								⚻+1.39	⚺+2.10	∠-1.64		□-4.13		+7.03	-6.48	+0.55
♄								P-1.60		P-3.20		□-3.18	-3.00	+10.00	-33.01	-23.01
♅									△+6.18			✶+5.47		+14.64	-14.77	-0.13
♆											∠-4.65	✶+3.65		+22.50	-5.13	+17.37
♇											✶+6.77	∠-1.53	-2.00	+26.75	-8.37	+18.38
MC														+18.52	-11.75	+6.77
ASC														+23.89	-14.46	+9.43
													Total	+211.30	-172.14	+39.16

Aquarian Age
Harmony/Discord Grill

Stellar Dynamics — Aquarian Age Chart

THE PLANETS, MIDHEAVEN AND ASCENDANT

ASTRODYNES		*HARMODYNES*		*DISCORDYNES*	
Mercury	78.67	Pluto	18.38		
Ascendant	65.30	Neptune	17.37		
Venus	57.71	Venus	11.48		
Sun	53.35	Mercury	10.78		
Uranus	52.44	Ascendant	9.43		
Pluto	49.23	Midheaven	6.77		
Saturn	46.96	Mars	3.57		
Midheaven	44.74	Jupiter	0.55		
Moon	35.87			Uranus	0.13
Jupiter	35.11			Sun	2.56
Neptune	32.87			Moon	3.47
Mars	11.46			Saturn	23.01
	763.71		78.33		29.17
				Net Harmony	49.16

THE SIGNS

ASTRODYNES		*HARMODYNES*		*DISCORDYNES*	
Virgo	127.64	Taurus	41.49		
Pisces	119.45	Pisces	22.73		
Capricorn	113.61	Cancer	7.69		
Taurus	110.96	Scorpio	5.49		
Aquarius	103.05	Libra	2.87		
Aries	84.94	Gemini	0.39		
Cancer	83.24	Sagittarius	0.28		
Leo	53.35			Leo	2.56
Gemini	39.34			Virgo	3.21
Sagittarius	17.56			Capricorn	7.16
Scorpio	15.17			Aquarius	14.13
Libra	14.43			Aries	21.57
	882.74		80.94		48.63
				Net Harmony	32.31

THE HOUSES

ASTRODYNES		*HARMODYNES*		*DISCORDYNES*	
Seven	155.50	Eleven	41.49		
Ten	146.68	One	7.69		
Three	114.98	Nine	5.70		
Eleven	110.96	Five	5.49		
One	83.24	Six	3.85		
Nine	82.56	Four	3.26		
Four	53.76	Twelve	0.39		
Twelve	39.34			Two	1.28
Six	29.02			Three	4.88
Two	26.68			Eight	5.79
Eight	24.85			Ten	10.32
Five	7.59			Seven	13.29
	875.16		67.87		35.56
				Net Harmony	32.31

THE AQUARIAN AGE

Unfortunately, there is no undisputed record of the date when the First of Aries among the constellations coincided with the First of Aries among the zodiacal signs, and there is no precise astronomical observation by which can be determined when the Aquarian Age began.

It would appear, however, that the first of Aries among the constellations must lie very close to the most brilliant and conspicuous star in the constellation Aries. This star, Alpha Arietis, now has a Right Ascension of 2 hours, 3 minutes. That is, the Vernal Equinox has moved back not quite 31 degrees since it was on the meridian occupied by this bright star in the head of the Ram. Calculating at the rate of precession by Right Ascension, in 1881 the Vernal Equinox was just about 30 degrees along the ecliptic west of the meridian of Alpha Arietis, so that if this star be used as a starting point for the circle of constellations, the Equinox backed into Aquarius just about 1881. .

In determining the time of the commencement of the Aquarian Age, as it is mere assumption that Alpha Arietis is the starting point in the circle of stars, we are faced with a problem similar to that of determining the hour of birth of a person when the precise time has not been recorded. It is essentially a problem of rectification, such as all astrologers are familiar with in their natal astrology practice.

Within the time limits which for other considerations seem reasonable, the most satisfactory method of rectifying a birth-chart is through the comparison of the events which have happened in the life with the positions found in the chart, and with the progressed aspects. And by the same token, rather than make calculations from Alpha Arietis, or from other equally uncertain starting points, it seems better to ascertain the commencement of the Aquarian Age from a consideration of events which clearly are not of the type which are characteristic of Pisces, through which for more than 2,000 years the Equinox moved by precession. The foregoing has been condensed from material to be found in "Spiritual Astrology," by C. C. Zain.

SOURCE: Chart was rectified by Elbert Benjamine and first published in the

DATE: Brotherhood of Light Annual for 1921.

THE AQUARIAN AGE

b. January 19, 1881, Washington, D.C., 38N53 77W01, 15:48:24 LMT

HISTORICAL EVENTS AND MAJOR PROGRESSIONS

EVENTS: Condensed from a listing given in the book "Astrological Lore of All Ages," by Elbert Benjamine.

1881—Venus p was semi-sextile Jupiter r, ruler of the President (10th), and opposition (separation) Uranus r, planet of sudden and unexpected events, ruler of the house of death (8th). James A. Garfield, 20th President of the U.S. was shot and killed.

1882—Mercury p, planet of intellectual interests, was sesqui-square Uranus r, planet of research and occultism. The Society for Psychical Research was founded.

1886—Jupiter p was inconjunct Uranus r, planet of revolution and agitation; and Mercury p, ruler of the house of antagonism (7th), was semi-square (friction) Venus p. Anarchist riots occurred in Chicago.

1888—Venus p, ruler of the house of diplomatic relations (9th), was semi-sextile (growth) Saturn r in the house of our Government (10th). The first Pan-American Congress met at Washington.

1890—Mars p was trine Neptune r. Neptune not only rules corporations but the watering of stocks and swindles. Congress passed the Sherman Antitrust Act.

1893—Mars p, ruler of the entertainment house (5th), was trine (luck) Uranus r, planet of invention. Edison developed moving picture apparatus. Mars p, ruler of the house of speculation (5th), also squared (obstacle) Jupiter r, a business planet in the business house (10th). There was a commercial panic.

1898—Sun p, co-ruler of the house of war (7th), was semi-sextile (growth) Mars p, ruler of war. Mercury p, chief ruler of the house of war (7th), was square (obstacle) Pluto r. As the result of war between the U.S. and Spain, Cuba was freed and the U.S. acquired Puerto Rico and the Philippines.

1907—Sun p, ruler of the money house (2nd), was square Pluto r, ruler of the speculation house (5th). A stock panic started in New York.

1914—Mars p, planet of war, was conjunction (prominence) Sun r, in the house of war (7th); and Sun p was sextile Mars r. World War I started.

1917—Sun p, co-ruler of the house of war (7th), was semi-square (friction) Saturn r, in the house of our Government (10th); and Mercury p sextiled Mercury r in the house of war (7th). The U.S. entered World War I.

1918—Mercury p, ruler of the house of war and partnership (7th), was trine Pluto r, planet of cooperation. World War I ended and a League of Nations formed to cooperate in preventing future wars.

1919—Venus p, ruler of the house of diplomatic relations and treaties (9th), was conjunction (prominence) Saturn p in the house of Government and the President (10th). Self-centered (Saturn) interest persuaded many senators to adopt an isolationist policy, and the Senate refused to ratify the League of Nations Covenant, thus insuring there would be another World War. President Wilson, on tour getting the people to insist the League of Nations Covenant be ratified, was stricken with a paralytic stroke, incapacitating him from usual presidential activities.

1929—Sun p, ruler of the money house (2nd), was opposition (separation) Moon r, ruler of the people (1st). A world-wide financial depression started.

1931—Mars p, planet of war, was inconjunct (expansion) Uranus p, the sudden planet. Japan suddenly invaded Manchuria, converted it into Manchukuo, placing that country and its resources under Japanese control.

1933—Mercury p was opposition (separation) Moon r, ruler of the people (1st); and Sun p, ruler of the money house (2nd), was semi-sextile Jupiter p, co-ruler of the Government (10th). As a result of both machinery and raw materials for production becoming concentrated in the hands of a few families interested primarily in profit, more than ten million people desirous of finding work were unemployed; so reducing purchasing power that banks all over America began to fail. President F. D. Roosevelt, when he took office in March, temporarily closed all the banks in the country. Then the Government arranged to insure deposits up to $5,000 and provide sufficient employment that no one need go hungry.

1934—Mercury p was still opposition Moon r (the people); and Mars p now came to the semi-sextile (growth), Venus r. The Western Powers, so engrossed in their own affairs, had been unwilling to bring heavy sanctions against Japan for acquiring Manchuria through military conquest. Japan now overran and annexed a slice of northern China.

1935—Sun p was sextile Pluto r, the gangster planet. The last of February and first of March, Mars t was moving slowly, turning retrograde. It was within close orb of opposition Uranus at this time, and made a perfect square to Pluto. Suddenly, on March 1, Hitler took over the Saar.

1936—Sun p was sextile Pluto r. As a result of Mussolini's success in using gangster tactics against a weaker nation, dictatorships gained tremendously in power and prestige.

1937—Sun p was sextile Pluto p; and Venus p, ruling the Ascendant (people) in the chart of Japan, was trine Uranus r. The Japanese moved to subjugate and take over the whole of China. They met unexpectedly stubborn resistance from the Chinese, and a war started that was to continue and later be an important part of World War II.

1938—Mars p came barely within orb of inconjunct (expansion) with Moon r; and Mercury p was conjunction (prominence) Venus r in the house of diplomatic relations (9th). Stalin, in March, conducted a blood purge in which many alleged traitors were executed. On March 12, 1938, Hitler invaded Austria.

1939—Mars p was inconjunct Moon r; and Mercury p, chief ruler of the house of war (7th), was opposition Uranus r. On March 15, Hitler invaded and subjugated Bohemia,

Moravia and Slovakia, accomplishing the conquest of Czechoslovakia. On March 28, Franco terminated the Spanish civil war, thus increasing the power of the totalitarian states. On April 1, Japan grabbed the French Spratly Islands. On April 6, Mussolini invaded and conquered Albania.

1940—Mars p was semi-square Mars r; Mercury p was opposition Uranus r; and Venus p was trine Uranus r, planet of the unusual. France and the Low Countries were overrun in May and conquered by Germany. On September 27, the Berlin-Rome-Tokyo Axis was formed. The U.S. traded obsolete destroyers to Britain for naval bases, a selective service act was passed to acquire soldiers, and F.D.R. shattered a precedent by being elected for a third term as president.

1941—Venus p, ruler of the house of treaties (9th), was conjunction Neptune p, ruler of Japan. Venus p was trine Uranus r. Mercury p opposed Uranus r, trined Venus r, and trined Neptune p, planet of idealism. Under these aspects the Big Three met in the Atlantic to agree on principles which were later to be called the Atlantic Charter. And on December 7, the Japanese attacked Pearl Harbor.

1942 — Mercury p opposed Uranus r and sextiled Neptune p. Through military conquests, Japan rapidly expanded her empire to vast proportions, conquering the Philippines, taking Singapore, Burma, the Dutch East Indies, New Guinea, and being stopped just short of invading Australia.

655M Fraternal Twin: [illegible]
(Name)

June 11 1917
(Month) (Day) (Year)

Place: Place not given

Latitude: 42 N 00

Longitude: 94 W 00

DOMINANT FACTOR

Time of Birth (Daylight Saving):

Correction for Standard Time:

Time of Birth (Standard Time):

Correction for Mean Time:

Local Mean Time of Birth, A.M or P.M.: 03:59:00

FIRST KEY PROBLEM

Noon: 11:60 12:00

Local Mean Time: 03:59

L.M.T. Interval: 08:01 Minus

Sidereal Time (Noon): 05:17:01

24:00:00

29:17:01

L.M.T. Interval: 08:01:00

S. T. (Uncorrected): 21:16:01

Correction, 9.86s per h. for E.G.M.T. Int.: :00:17

Sidereal Time (Of Birth): 21:55:44

SECOND KEY PROBLEM

Local Mean Time of Birth: 3:59 a.m.

Hrs. and mins. E. or W. of Greenwich: 6:16 W

E.G.M.T.: 10:15 a.m.

Noon: 11:60 12:00

E.G.M.T.: 10:15

E.G.M.T. Interval (Indicate plus or minus): 1:45 Minus

Dalton's Tables of Houses
Die Deutsche Ephemeris

ADDITIONAL FACTORS

Constant Log: 1.1372

Limiting Date (Including year): July 7, 1917

Doris Chase Doane

16♒28
23♑56
3♑56
12♐03
27♎22
16♍10
16♌28
23♋56
3♋56
12♊03
27♈22
16♓10
♅ 23♒39 ℞
☽ 8♓40
♃ 25♉57
☿ 26♉39
♂ 27♉33
☉ 19♊54
♀ 2♋13
♇ 3♋29
♄ 28♋30
♆ 3♌01

Mov.	Fix.	Mut.	Fire	Earth	Air	Water	Ang.	Suc.	Cad.
3	M.C. 5	Asc. 2	1	3	M.C. 2 Asc	4	5	0	5
Per.	**Comp.**	**Pub.**	**Life**	**Wealth**	**Assoc.**	**Psy.**	**Above**	**East**	**Ret.**
8	0	2	3	2	2	3	5	10	1

ASPECTS

Declinations		☉	☽	☿	♀	♂	♃	♄	♅	♆	♇	MC	ASC
23 N 04	☉	.	□	.	☌	.	.	.	△	∠	☌	△	☌ P
4 S 11	☽			.	△	□	.	⚼	.	.	△	.	□
15 N 53	☿				.	☌	☌	✶	□	.	.	. P	.
24 N 13	♀					.	.	.	△	⚺	☌	⚼	.
19 N 29	♂						☌	✶	□	. P	. P	.	.
18 N 27	♃							✶	□	. P	. P	□	.
20 N 51	♄								.	☌	.	.	∠
14 S 20	♅									.	△	☌	.
19 N 19	♆										⚺ P	.	.
18 N 51	♇											⚼	.
15 S 54	M.C.												△
22 N 15	Asc.												

666F Fraternal Twin: SUN
(Name)

June 11 1917
(Month) (Day) (Year)

Place: Place not given

Latitude: 42 N 00

Longitude: 94 W 00

DOMINANT FACTOR

Time of Birth (Daylight Saving):

Correction for Standard Time:

Time of Birth (Standard Time):

Correction for Mean Time:

Local Mean Time of Birth, A.M or P.M.: 04:14:00

FIRST KEY PROBLEM

Noon: 11:60 12:00

Local Mean Time: 04:14

L.M.T. Interval: 07:46 Minus

Sidereal Time (Noon): 05:17:01

24:00:00

29:17:01

L.M.T. Interval: 07:46:00

S. T. (Uncorrected): 21:31:01

Correction, 9.86s per h. for E.G.M.T. Int.: 00:15

Sidereal Time (Of Birth): 21:30:46

SECOND KEY PROBLEM

Local Mean Time of Birth: 4:14 a.m.

Hrs. and mins. E. or W. of Greenwich: 6:16 W

E.G.M.T.: 10:30 a.m.

Noon: 11:60 12:00

E.G.M.T.: 10:30

E.G.M.T. Interval (Indicate plus or minus): 1:30 Minus

Dalton's Tables of Houses

Die Deutsche Ephemeris

ADDITIONAL FACTORS

Constant Log: 1.2041

Limiting Date (Including year): July 4, 1917

Doris Chase Doane

20♒17
20♓51
2♉28 (♈)
16♊03
27♑19
7♐15
16♐03
2♏28 (♎)
20♍51
20♌17
27♋19
7♋15
♅ 23♒39 ℞
☽ 8♓48
♃ 25♉57
☿ 26♉40
♂ 27♉33
☉ 19♊54
♀ 2♋13
♇ 3♋29
♄ 28♋30
♆ 3♌01

Mov.	Fix.	Mut.	Fire	Earth	Air	Water	Ang.	Suc.	Cad.
3	M.C 5	Asc. 2	1	3	M.C, Asc. 2	4	5	0	5
Per.	**Comp.**	**Pug.**	**Life**	**Wealth**	**Assoc.**	**Psy.**	**Above**	**East**	**Ret.**
8	0	2	3	2	2	3	5	10	1

Declinations	ASPECTS	☉	☽	☿	♀	♂	♃	♄	♅	♆	♇	MC	ASC
23 N 04	☉		□	·	☌	·	·	·	△	∠	☌	△	☍P
4 S 08	☽			·	△	□	·	Q	·	·	△	·	□
15 N 53	☿				·	☌	☌	✱	□	·	·	□	·
24 N 13	♀					·	·	·	△	⚺	☌	Q	·
19 N 29	♂						☌	✱	□	·P	·P	□	·
18 N 27	♃							✱	□	·P	·P	□	·
20 N 51	♄								·	☌	·	·	∠
14 S 20	♅									·	△	☌P	△
19 N 19	♆										⚺P	·	∠
18 N 51	♇											Q	·
14 S 25	M.C.												△
22 N 44	Asc.												

667M Astrological Twin: JOE
(Name)

May 27 1883
(Month) (Day) (Year)

Place: Place not given

Latitude: 42 N 00

Longitude: 75 W 00

DOMINANT FACTOR

Time of Birth (Daylight Saving):

Correction for Standard Time:

Time of Birth (Standard Time):

Correction for Mean Time:

Local Mean Time of Birth. A.M or P.M.: 08:54:00

FIRST KEY PROBLEM

Noon: 11:60 12:00

Local Mean Time: 08:54

L.M.T. Interval: 03:06 Minus

Sidereal Time (Noon): 04:18:48

L.M.T. Interval: 03:06:00

S. T. (Uncorrected): 01:12:48

Correction, 9.86s per h. for E.G.M.T. Int.: 00:19

Sidereal Time (Of Birth): 01:13:07

SECOND KEY PROBLEM

Local Mean Time of Birth: 8:45 a.m.

Hrs. and mins. E. or W. of Greenwich: 5:00 W

E.G.M.T.: 13:54

Noon: 12:00 12:00

E.G.M.T.: 1:54 p.m.

E.G.M.T. Interval (Indicate plus or minus): 1:54 Plus

Dalton's Tables of Houses
Die Deutsche Ephemeris

ADDITIONAL FACTORS

Constant Log: 1.1015

Limiting Date (Including year): April 28, 1883

Doris Chase Doane

19♈48
18♓43
24♒15
4♒14
3♑19
♐
26♏48
19♎48
18♍43
24♌15
4♌14
3♋19
♊
26♉48
♂ 28♈03
♀ 5♉19
♆ 19♉12
♇ 29♉47
♄ 0♊23
☉ 5♊54
☿ 21♊19℞
♃ 4♋32
☽ 10♒14
♅ 19♍13 Sta.

Mov.	Fix.	Mut.	Fire	Earth	Air	Water	Ang.	Suc.	Cad.
M.C. 2	Asc 3	5	M.C 1 Asc	4	4	1	4	4	2
Per.	**Comp.**	**Pug.**	**Life**	**Wealth**	**Assoc.**	**Psy.**	**Above**	**East**	**Ret.**
2	1	7	0	3	6	1	9	9	1

Declinations		☉	☽	☿	♀	♂	♃	♄	♅	♆	♇	MC	ASC
21 N 17	☉		△	·	⊻	·	⊻	☌	·	·	☌	∠	✱
12 S 37	☽			△	□	·	·	△	·	□	△	·	☍
23 N 22	☿				∠	✱	·P	·	□	⊻	·	✱	∠
11 N 37	♀					☌	✱	·	Q	·	·	·	□
9 N 55	♂						✱	⊻	·	·	⊻	☌	□
23 N 23	♃							·	·	∠	·	·	⊻
18 N 29	♄								·	☌	☌	∠	✱P
4 N 59	♅									△	:	⊼	∠
15 N 54	♆										☌	⊻	·
7 N 33	♇											·	✱P
7 N 45	M.C.												·
19 N 12	Asc.												

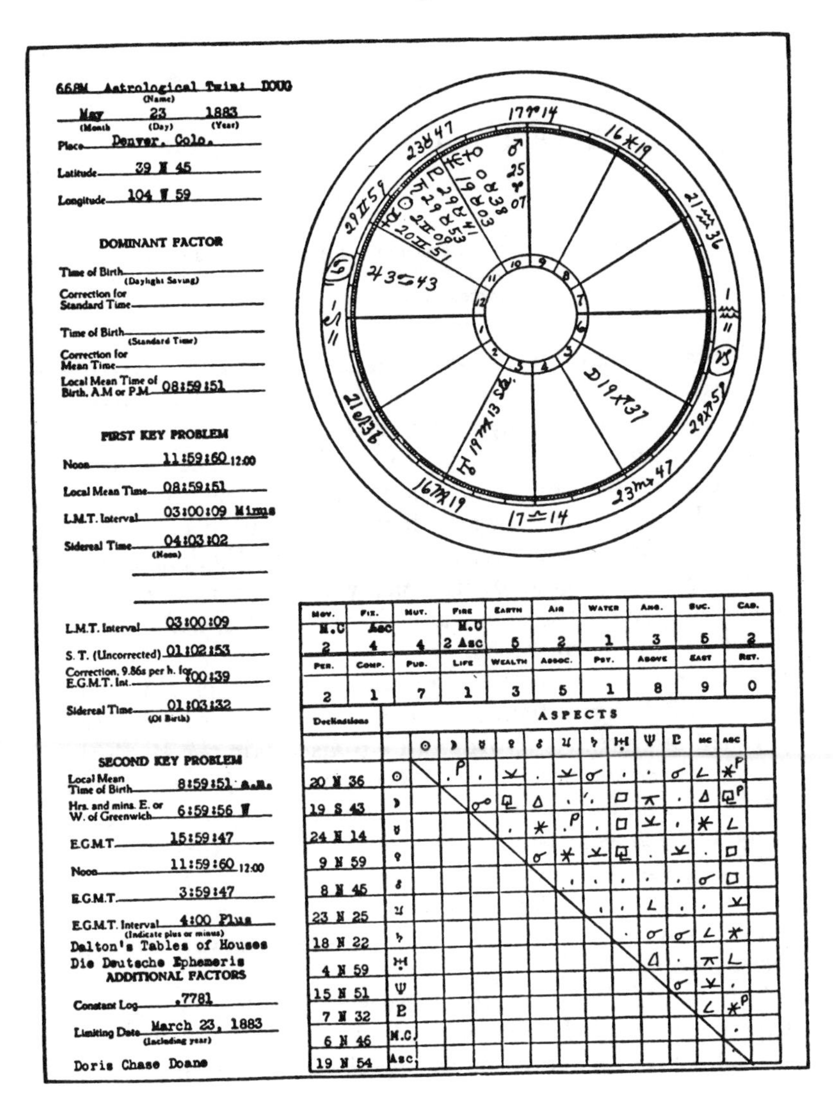

668M Astrological Twin: DOUG
(Name)

May 23 1883
(Month) (Day) (Year)

Place Denver, Colo.

Latitude 39 N 45

Longitude 104 W 59

DOMINANT FACTOR

Time of Birth (Daylight Saving) ______

Correction for Standard Time ______

Time of Birth (Standard Time) ______

Correction for Mean Time ______

Local Mean Time of Birth, A.M or P.M 08:59:51

FIRST KEY PROBLEM

Noon 11:59:60 12:00

Local Mean Time 08:59:51

L.M.T. Interval 03:00:09 Minus

Sidereal Time (Noon) 04:03:02

L.M.T. Interval 03:00:09

S. T. (Uncorrected) 01:02:53

Correction, 9.86s per h. for E.G.M.T. Int. 00:39

Sidereal Time (Of Birth) 01:03:32

SECOND KEY PROBLEM

Local Mean Time of Birth 8:59:51 a.m.

Hrs. and mins. E. or W. of Greenwich 6:59:56 W

E.G.M.T. 15:59:47

Noon 11:59:60 12:00

E.G.M.T. 3:59:47

E.G.M.T. Interval (Indicate plus or minus) 4:00 Plus

Dalton's Tables of Houses

Die Deutsche Ephemeris

ADDITIONAL FACTORS

Constant Log .7781

Limiting Date (Including year) March 23, 1883

Doris Chase Doane

Mov.	Fix.	Mut.	Fire	Earth	Air	Water	Ang.	Suc.	Cad.
M.C 2	Asc 4	4	M.C 2 Asc	5	2	1	3	5	2
Per.	**Comp.**	**Pub.**	**Life**	**Wealth**	**Assoc.**	**Psy.**	**Above**	**East**	**Ret.**
2	1	7	1	3	5	1	8	9	0

ASPECTS

Declinations		☉	☽	☿	♀	♂	♃	♄	♅	♆	♇	MC	ASC
20 N 36	☉		. P	.	⊻	.	⊻	☌	.	.	☌	∠	⚹ P
19 S 43	☽			☍	⚼	△	.	'.	□	⚻	.	△	⚼ P
24 N 14	☿				.	⚹	. P	.	□	⊻	.	⚹	∠
9 N 59	♀					☌	⚹	⊻	⚼	.	⊻	.	□
8 N 45	♂						.	.	.	.	.	☌	□
23 N 25	♃							.	.	∠	.	.	⊻
18 N 22	♄								.	☌	☌	∠	⚹
4 N 59	♅									△	.	⚻	∠
15 N 51	♆										☌	⊻	.
7 N 32	♇											∠	⚹ P
6 N 46	M.C.												.
19 N 54	Asc.												

VIII
CASE HISTORY STUDIES

"The most effective method of becoming skilled in astrology is the case method, by which the student, after gaining some knowledge of theory, studies actual cases in which astrological positions have coincided with recorded events or conditions."—C. C. Zain.

For more than thirty years The Church of Light Astrological Research activities have been carried on by trained volunteer workers, headed by Elbert Benjamine from April 1924 to November 1951, and by Doris Chase Doane since then. This cooperative effort probing many phases of life was founded to learn just how astrology could be used as an effective tool for living a completely constructive life. Finding a method to approach this goal offered many complex problems. But as will be demonstrated in case history studies, some of these problems have been solved.

The first phase of astrological research was objective in nature, involving statistical studies and analyses of large series of timed birth-charts in order (1) to ascertain the Birth-chart Constants for as many vocations as possible, (2) to ascertain the Birth-chart Constants and Progressed Constants for as many diseases as possible, and (3) to ascertain the Birth-chart and Progressed Constants for each type of event which may enter human life. This work is still carried on—one lifetime would not provide the time and energies to cover these categories completely.

But in addition to these factors, other highly important specialized knowledge was needed. The next phase of research was subjective in nature, involving the study of thousands of case histories relating to timed birth-charts and their progressions in order to solve FOUR problems:

I—Find the factors, at the time of a given progressed aspect, that determine the Specific Event which then takes place.

II—Find the factors that, with a given birth-chart and a given set of progressed aspects, determine whether or not an individual can succeed in accomplishing some definite thing, in spite of difficulty and stress.

III—Find the factors, when with a given birth-chart and given progressed aspects the attracted event is much more or much less important than normally expected, that influenced the event to depart in importance from the norm.

IV—Find the factors, when with a given birth-chart and given progressed aspects the attracted event is much more harmonious or much more discordant than normally expected, that influenced the event to depart in harmony or discord from the norm.

Years of research resulted in solving these problems. The solutions provide a wealth of information to be found in the 21 Brotherhood of Light Courses.

Another reason for spurring research activity was the prevailing wide-spread practice of confusing ESP precognition principles with the principles of astrology—a confusion hindering the acceptance of scientific astrology.

Through astrology or ESP, or both, life is constantly being influenced by two different types of environment. Man reacts to inner-plane forces and to outer-plane conditions. Coincident with progressed aspects of a given kind, events typical of the planets involved enter the life. On the physical side, our reactions to people and the weather influence our conduct and the events which arrive.

Materialists assume that man reacts only to physical forces, and some astrological enthusiasts assume every event from the "cradle to the grave" takes place solely because of the planets. Research amply proves that neither of these extreme views is warranted, and that the events entering an individual's life and the conditions surrounding his physical existence at any time, are the products of his character at that time as energized by progressed aspects acting upon and reacting from his outer-plane environment.

Anything seen in the future through ESP is usually viewed as a specific event, in definite detailed surroundings. The complete picture may be verified when it comes to pass. Looking on the astral plane at the projected world-lines makes it possible for the imagination to picture a future event with all its ramifications, and later its replica is actually seen on the physical plane.

However, natal astrology is an entirely different phase of occult study. Instead of looking down world-lines to project the consciousness into the future, natal astrology embraces (1) inspecting character forces mapped by birth-chart planetary positions; (2) considering what environmental forces have impinged upon and modified that character in the past; and (3) what environmental forces will stimulate the modified character at a certain period in the future. Studying the interaction of forces between character and environment makes possible the estimation of what conditions and events will enter the life during a given period.

In natal astrology events are not seen. If a specific event is predicted, a fortune is being told—a process different than that used in scientific astrology. In scientific astrology, the basic character tendencies (birth-chart) and environmental forces (progressions) interact according to conditioning (case history); the result enables a deduction of what will happen if nothing is done about it. Even with this information, specific events cannot be predicted: only a predisposition is foreshown.

If a man and a dog had practically identical birth-charts and progressed aspects, the events mapped in the dog's life would not necessarily be the same as those in a man's life. Therefore, the exact nature of what will happen must be judged from the evolutionary level.

Judging the probable events by natal astrology, whether in the life of an insect, a dog, or a man, conclusions are not reached through ESP precognition but through a careful weighing of the probable reaction of character to forces—past, present and future. Resistance to certain specific events in a given category may be great, and resistance to other specific events in the same category may be small. Or the resistance to all events in a certain category may be very great or very small.

Researching for methods to control life, and live it to the best, required knowledge of how much impact each factor influencing life had in attracting events and molding destiny.

In all problems relating to events which come into the life, seven factors were found which, in differing degrees, appeared to cause a difference. For the discussion of these factors and problems arising out of them, the following Astro-Physical Interaction diagram was adopted:

ASTRO-PHYSICAL INTERACTION DIAGRAM

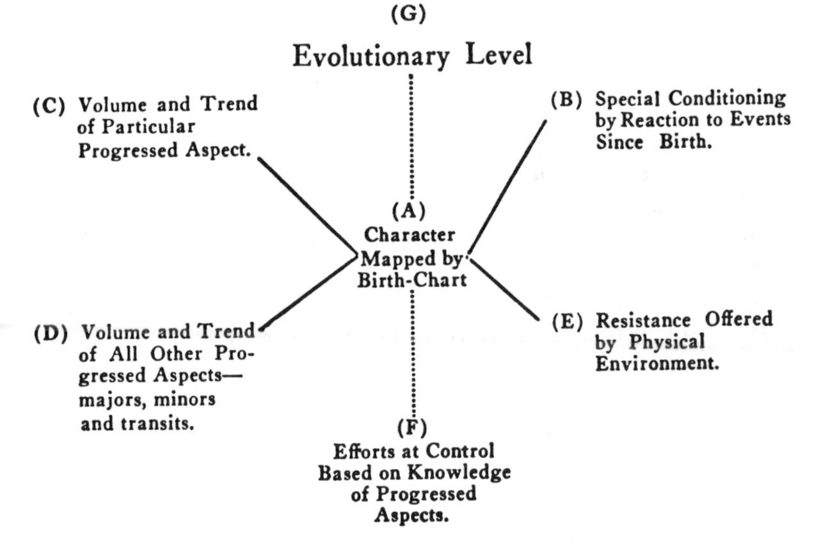

For convenient discussion each letter of the factors indicated above will be used to save space. Other abbreviations: *p*, major progressed position; *m*, minor progressed position; *t*, transit progressed position. Numbers in parentheses preceded by # indicate constants for a definite event, disease or vocation, designated by the number. These will be found in the section on Astrological Research Reports; e.g., (#1) designates the Movie Actor Report.

(B) and (E) are mainly controlled by parents raising children, and may largely be brought under control of a mature individual if he is willing to make sufficient effort. It is important to know how much influence these two outer-plane factors ordinarily have in comparison to that exerted by the two inner-plane factors (C) and (D). As practical effort, based upon astrological knowledge, yields valuable results, the study of these five factors provided information upon which to base control techniques.

Case histories bearing upon this and other problems were studied. The method of research can best be demonstrated with two closely similar birth-charts accompanied by case histories containing date and nature of events and special conditioning. The evolutionary level, even in the case of twins, may be different, but these studies will show the effect of different conditioning, and the effect of widely different types of environment at the time a given progressed aspect is operative.

Case history studies of the lives of identical twins, natural twins and astrological twins proved that harmony or discord built into the thought-cells of any of the ten types of mental urges can be altered through training (conditioning). That is, one born with harmonious thought-cells of a given type can train them to feel, and strive to express, discordantly; and one born with discordant thought-cells of a given type can train them to feel, and strive to express, harmoniously. This process, whether deliberate or unconscious, through response to environmental experiences is called reconditioning. All persons are to some extent reconditioned after birth—not only by harmony and discord, but by the specific things for which their thought-cells work.

Astrological twins are those born to different parents on the same day, or those having nearly identical birth-charts, but not born on the same day. Charts of astrological twins proved more instructive in our research than the charts of twins born of the same parents, because physical environmental factors were more divergent.

To demonstrate how physical environment modifies stimulation mapped by progressed aspects, the charts of twins born of the same parents will be used. Twins may be one of two different types: identical twins develop from one fertilized ovum, are always of the same sex and have similar appearance; fraternal twins develop from two fertilized ova, may be of either sex, and may or may not look alike.

As one is male and the other female, these twins are fraternal and not identical. For convenience, they have been fictitiously named Stan and Sue. Stan was born June 11, 1917, 42N00, 94W00, at 03:59:00 L.M.T. Sue was born on the same day at the same place, but the hour was 04:14:00 L.M.T. Their births were separated by fifteen minutes. The following study covers about twenty years of their lives.

FRATERNAL TWINS

If sex is an integral part of the character, there is a real difference in factor (A), causing divergent thought-cell reactions both to progressed aspects and to outer-plane environment. Even with birth-charts otherwise identical, difference in sex implies essential difference in the endocrine glands, which pronouncedly influence a person's reaction to situations.

Due to glandular makeup, for instance, a girl commonly matures several years earlier than a boy. Consequently, a progressed aspect occurring in the early 'teens in both charts might map a juvenile reaction from the boy and a mature reaction from the girl. But of even greater significance would be the influence of the opposite sex on each life; in a male chart the opposite sex is indicated by the Moon, while in a female chart, by the Sun. In the case of twins of opposite sex with closely similar charts, astrological conditions alone indicate that their general experience with partnerships will be similar, but their experiences in love and marriage will be dissimilar.

In these charts the difference in glandular makeup and modification of character through conditioning resulted in varied dispositions, ambitions and experiences with the opposite sex.

The twins are similar in physical build; both large with barrel chests. Sue is blond, with china blue eyes; Stan has somewhat darker skin and hair, with brown eyes.

Apart from glandular reactions due to sex, twins reared in the same household are subject to a special conditioning arising from constant association with each other. Decisions must be made on matters of mutual interest. Time and energy are often limited for extensive parleys in which both have an equal voice. Consequently, one of the pair gradually becomes conditioned to exercise leadership and the other to be led. It has been noted at twin conventions that one twin gives a positive impression and the other negative, even when they are markedly similar in appearance.

This female twin chart has the Sun about 4 degrees closer to the exact conjunction aspect with the Ascendant, and Uranus about 4 degrees closer to the exact conjunction with the Midheaven. As closer aspects represent a greater volume of energy, the female might be expected to exert more energy and aggression than the male.

Planets in intercepted signs represent qualities needing considerable external stimulation to express as fully as if not intercepted. In the twin charts, three planets not intercepted in the female chart are intercepted in Taurus in the male chart.

Whether attributed to difference in glandular makeup due to sex, the difference in planetary positions, or somewhat to both, the fact remains that—at least—up to 20 years of age, the female was the positive and the male was the negative.

Undoubtedly this difference, as well as other differences in character, increased through early environmental conditioning. Probably the most potent factor in this different conditioning was that the female, an only girl with three brothers, was raised chiefly by the mother, a widow.

Being the only girl, Sue was pampered, and the boys were always compelled to yield to her. They developed the attitude that it was up to them to provide what she wanted; and she developed the attitude that she should thus be provided for. The boys found that to get anything they wanted they had to work for it. The girl found that to get what she desired she had to work the boys for it. Tenth-house Uranus trine Sun denotes a headstrong individual. As Uranus is closer to the cusp of her 10th, she probably—from astrological conditions alone—was more headstrong than her twin brother or the balance of the family.

From childhood she enjoyed rough play. Always ready to go fishing or hunting with her older brothers, she was not reticent about suggesting such activities. She played baseball, and won most of the marbles from school playmates. Later she became a star on the soft ball and basket ball teams, and before finishing high school her tenth-house Moon mapped the statewide reputation she gained in basket ball contests. She was an extrovert (Sun in 1st house) who gained a sense of significance through athletic (#75)[1] superiority.

When children having a strong drive for significance associate with other children who outclass them in performance, they usually feel inferior. There may be a tendency for them to seek some other channel of expression for their energies—a channel where without unbeatable competition they excel and gain the feeling of significance they strongly crave.

Even though Stan was powerful and could have been an athlete if he wanted to, early experiences with his sister discouraged the belief he could gain significance through such endeavors. She was always willing to fight and, having been pampered, would use a rock or a club to gain the advantage. Had he—as a youngster—given her a beating, the action would have been too costly for his older brothers would have taken vengeance. Besides, he was always tremendously fond of her.

Like his sister, he has one negative planet in the 10th and two negative planets in the 1st. He reacted sensitively to his brothers and tomboy sister, becoming the most introvertive member of the family.

He avoided hunting, fishing, athletics and strenuous games, always finding himself outclassed by the others. Loving animals, he spent a lot of time with his pets, and raised chickens, ducks, goats, pigs, calves, horses, sheep, rabbits, and white mice. Finding he could sell the animals he raised to contribute to the

---

[1]Number preceded by # designates Astrological Research Report.

needy family budget (his mother was widowed early), he gained a sense of significance encouraging him to make further monetary efforts.

When he made purchases for the family or some member of it, he noticed that he rose in the family's esteem. This rise in station satisfied his strong drive for significance, mapped by the rising Sun, that his sister satisfied by athletic leadership. In the matter of studies also, the competition of his sister proved too strenuous and he dropped behind her in school at the third grade; when in both charts Sun p[2] reached semi-sextile Saturn r[3]. At the time, she had Ascendant p trine Uranus r, which he did not have until four years later.

The girl being ahead of him in school seemed to increase his shyness and reticence. As his brothers left home, the weight of financial responsibility grew heavier upon him, and his sense of significance increased in being the provider for the family. Sue's responsibilities increased also; she was assigned more chores about the home. Saturn is in the house of brethren (3rd), and one brother left home in 1926 when in both twin charts Sun p was semi-sextile Saturn r, and Venus p was semi-square Jupiter p. The older brother left in 1927, when Sun p was semi-sextile Saturn p, and other aspects shortly to be mentioned were operative.

1927, however, brought another event which was more serious to the sister. Probably due to difference in glandular makeup because of sex, and environmental conditioning, she lived far more strenuously than her brother. She rushed into things pell-mell. Jupiter p conjunction Mars r increased thought-cell activity manifesting as general excitability, and Mercury p made the square of Moon r. The Moon maps certain more pronounced influences over the physiological functions in the life of a girl, and even at ten years of age an affliction to it by progression may be more serious in the chart of a girl than in a boy's chart.

Observation of progressed aspects in other charts indicates

---

[2]p, major progressed position.

---

[3]r, birth-chart position.

that one additional progressed aspect often maps greater intensity to the significance of what happened. Until more studies are made of such case histories, however, it is premature to decide how much should be attributed to glandular difference due to sex, how much to environmental conditioning, and how much to the addition of one progressed aspect at the time. In the girl's chart, and not in the boy's, Ascendant p was semi-sextile Jupiter r. She began to have heart difficulties (#33) which later were increased through strenuous athletics and immoderate smoking. The boy suffered no heart trouble.

Mars is the ruler of work (6th) and the Moon rules business (10th) in these charts. In 1933 Mars p came to the square of Moon r, and the twins went to work (#52). Stan had a little temporary work in a store, but the only steady job he could get was working on a farm where the hours were long and the pay small. Nevertheless, he gained a feeling of significance, unattainable in school, by bringing revenue to the family. As a consequence, under this aspect he quit high school in his sophomore year to devote his energies to money-making.

He could always get a job on a farm, but times were hard. Mars—ruling both work and animals (6th and 12th)—afflicted the ruler of his money house (Moon) mapping difficulty collecting his wages in cash. Finally (note three planets in 12th—animals) he collected an assortment of pigs, calves, colts and poultry in lieu of wages due (#15), raised them and later on converted them into money. This initial venture into business conditioned him still further to find significance through such lines of endeavor.

There were no farm opportunities for a girl. Yet Mars p square Moon r, ruling the house of money (2nd), mapped a powerful urge to acquire cash (#14). The only thing open for a girl was cafe work in town, where not long hours but late hours at night were required. Many customers started flirtations with any girl they could.

Sue worked in the cafe only during vacations until she graduated from high school. To relinquish her studies, as the boy had done, would have meant also to relinquish other things in which she found her greatest satisfaction. Not only did Sue get along better with her schooling than Stan, but gaining her greatest sense of significance through athletic participation required her enrollment at school.

In 1934 Stan had Midheaven p trine Venus r and Pluto r. Sue had these progressed aspects when she was about 13 years of age and therefore possibly when they were less effective. Mars p was still square Moon r; Mercury p had moved up to conjunction Sun r; and Venus p had reached the inconjunct with Uranus r.

As the Sun rules the opposite sex in a girl's chart, it is to be expected that Mercury p (ruler of the natal Ascendant) coming to the conjunction (prominence) aspect with Sun r would map in some manner the opposite sex prominently entering her life. Sun trine Uranus in her birth-chart, as well as Venus p inconjunct Uranus r, shows that she would not be unduly restrained by conventions.

While these Venus progressions were operative, Stan became fairly fond of a girl—fond enough that when he left the region in late 1935 he carried her photo with him and kept it in a prominent place in his room. But he was not fond enough to ever write her.

He so thoroughly disapproved of his sister's care-free ways and wild tendencies that it seemed to condition him to an equal degree in the opposite direction. Her cafe associations led her to think nothing of taking a drink. She became a heavy smoker, to the detriment of her heart. And the conventions did not bother her overly much. If she did not arrive home soon after the cafe closed, the boy paced the floor and worried until she put in an appearance. He did not smoke, and only after Mars p moved into his 1st house, late 1937, could he be induced to take so much as a glass of beer. Furthermore, his attitude toward women in general was beyond reproach.

Working away from town in 1934 and looking after his accumulating livestock when not busy for others, furnished Stan a different environment than his sister had. Even had he the inclination, it would have been difficult for him to work as he did and at the same time live the sort of life as his twin. Different contacts were offered Sue by the cafe patrons—one of whom was the son of wealthy parents living in another town. He had much more money than is good for any young man to spend, and arriving in his showy car, he would meet Sue after work and drive to the nearest city, where they would visit after-hours night clubs, he squandering large amounts of cash.

This companionship which started under Venus p inconjunct Uranus r, and Mercury p conjunction Sun r in 1934 developed an even more serious turn for the girl in 1935, when Ascendant p came within orb of conjunction Venus r. Not only did the recurrent good times result in complications, but Sue fell deeply in love with her wealthy boy friend.

The mother of the twins is accurately described by Uranus in the 10th house of their charts. But—due to sex or to difference in conditioning—they developed an opposite attitude toward her. She always leaned strongly on Stan and praised him for his steady qualities. He experienced pleasure in doing things for her. Not only did he provide for her financially, but presented flowers on her birthday and treated her like a sweetheart.

But the mother spoiled Sue early through letting her take advantage of her brothers and have her own way. And while the boy appeared never to notice any faults of his mother, Sue resented the mother's Uranian peculiarities and was highly antagonistic toward her. Mother and daughter quarreled violently and often.

1935 brought Mercury p trine Uranus p. Stan, finding business opportunities too poor in the part of the country where he lived, decided to go elsewhere. He worked at a farm all summer and fall. He had never had a car, nor learned to drive one. And the wages he received were small. But Sue, who at the time had Ascendant p conjunction Venus r, which he did not have, talked him into buying her a car so she could more easily go to work.

Jupiter p in both charts was sextile Saturn p, ruler of the

house of long journeys (9th) and in the house of short journeys (3rd). By late fall Sun p, ruler of the home (4th), was within orb of sesqui-square Uranus p. Stan moved across the country to a distant city (#12), arriving there on Thanksgiving Day. Moon p at the time was trine Pluto r and square Neptune r.

Even though Mercury p was trine Uranus r, the sesqui-square of Sun p to Uranus r mapped difficulty in getting employment, and Stan picked up such odd jobs as he could until mid-summer 1936. If Sue was having difficulties they were of another kind, mapped by Ascendant p conjunction Venus r, and related to her boy friend.

Hitherto Stan had disliked construction work, or Mars work of any kind. But by mid-summer 1936 Mars p had come within orb of conjunction to Ascendant r. However, Sue did have Ascendant p conjunction Venus r which Stan did not. Both had Sun p (ruler of the home) trine Moon r; Sun p sesqui-square Uranus p; Venus p sextile Jupiter r; and Venus p semi-square Mars p.

Stan had an opportunity to leave the city on a construction job for good wages at a rather isolated spot. As soon as he was located and had his first paycheck he sent for his mother and sister. He had made a long journey (#12) the previous Thanksgiving which his sister did not make, although she took shorter trips in her new car. To reach his job he took a short trip (#13) in 1936, while in July Sue took a long journey to reach the same place. Moon p was inconjunct Mars p.

The change, mapped by Sun p sesqui-square Uranus r, was as radical for one as for the other—it meant leaving behind all old conditions and entering an entirely new environment for the future. It brought Stan association with roughneck construction workers and a new type of work. And it broke off whatever companionship there had been between himself and the girl he had at least cared something about. But for Sue social and affectional relations were drastically altered.

Under Ascendant p conjunction Venus r she had been having altogether too good a time. Her boy friend had spent much in giving her pleasure. She had become deeply involved with him, but he seemed disinclined to marry her. When she moved, it not only brought separation from him, but at the location of the construction camp there was not even a movie.

About Thanksgiving, 1936, the job played out (#53) and all three moved back to the city (#12). Stan, however, finding he could do construction work, immediately got a carpenter's job (#52) and was able to maintain his mother and sister through fairly steady work at good wages (#14).

In the lives of both brother and sister agitation (mapped by Sun p sesqui-square Uranus r) and friction (Venus p semi-square Mars p) were conspicuous during late 1936 and much of 1937. Sun p trine Moon r, and Venus p sextile Jupiter r, then sextile Mercury r, coincided with steady work and sufficient income for Stan. Sue's discontent and actions not only kept him upset, but cost him more money (#15) than he could afford.

A progressed Sun-Uranus aspect in any chart maps new people strongly entering the life, but in a female chart—because the Sun rules the opposite sex—this aspect maps a strong attraction of men into the environment. Bringing new acquaintances, the environmental change broke old ties for both twins. And possibly to show her old boy friend she could marry if she wanted to, Sue decided impulsively to marry a recent acquaintance.

To point up the difference in the twins' characters here, when Stan went among people, especially when dating a girl, he was usually careful of dress and personal appearance. He dolled himself up, using ample perfume; in fact, was a bit vain of his appearance. In this he seemed more effeminate than Sue who was less careful of her appearance than the average young woman of her age.

Both were impetuous about different things. Sue was impetuous where the opposite sex was concerned. But the Moon in Stan's chart does not aspect Uranus, and his association or experience with the opposite sex had been, up to 20 years of age, quite limited. Early in life he found he could gratify his desire for significance through financial channels. His powerful incentive to make money and use it in such a way as would bring him esteem of others, not only was a dominant motive, but when he got money he was as reckless in its expenditure as Sue was reckless in other ways.

He never refused an acquaintance a loan if he had the cash, even when he knew he would never get it back. When he went any place with his work crew, he insisted on paying all the bills for the entertainment. When he dated a girl, a mere movie was not good enough; he got expensive tickets to the opera, and afterwards had an expensive dinner at a swank eatery. In various ways he used money in a manner that most people think careless to attain a sense of significance. His sister became quite expert, on the other hand, in finding ways to get others to pay her bills and seemed to gain a sense of significance from it.

It is the usual thing for an unmarried woman under the aspects of progressed Venus, and Sun p afflicting Uranus r, to act impulsively where attachments with the opposite sex are concerned. May 7, 1937, just as Moon p had reached its one degree orb applying to square of Uranus p, Sue was married (#2).

Stan did not get married; perhaps it is an understatement to say he did get the full force of the stimulation mapped by the aspect. His sister and her husband moved in with him and his mother. Having no financial resources of their own, he purchased their furniture and otherwise set them up in housekeeping. Furthermore, the newly married couple began quarreling the night after they were married, and thereafter the household was in an uproar.

Probably Sue's affections were still with her wealthy boy friend in another part of the country. No doubt after being the companion of a man who lavished money showing her a good time, it was too great a change to settle down with another man without an income to spend, and who had to borrow from her brother. Likely, also, under the progressed aspect to Uranus, she had made a poor choice. The upshot of the matter was after two weeks of unhappily married life she left her husband (#10) abruptly and returned (#12) to the section of the country where her previous boy friend lived, greatly to the agitation of her brother.

Both the twins were fond of music and sang well. Stan had been told he had possibilities if he would train his voice (#47). In the city where he worked, friends encouraged him to take vocal lessons. Not only was he given an additional sense of significance through his singing; he was told that if he made good it would mean much money.

With visions of future wealth and fame, he took vocal lessons and studied music during 1937, 1938 and part of 1939. Jupiter p (ruler of the 7th) sextile Saturn r (ruler of the house of studies—3rd), and Venus p sextile Mercury r (general ruler of mental activity) mapped neither a marriage nor a long journey in 1937; nor did Venus p conjunction Saturn r, in 1938, 1939. Instead, he engaged in hard study. Previous to this time he had never been much of a student; but under the progressed aspect of Mercury and to the ruler of the 3rd house, when he found it necessary to gain significance, he could study as hard—he studied Latin among other things—and make as good progress as his sister had done while she was in high school.

After leaving her husband and while on the trip to her earlier environment, Sue was in an auto accident (#37) on May 22. She was thrown through the windshield, badly cutting her face and knees. Moon p was square Uranus r at the time she left her husband and when the accident occurred. Whatever her ideas were in making this trip, it appears they did not turn out well, for when Moon p cleared the Uranus p aspect, about a month after she left, she returned to her husband to try marriage again. This effort at reconciliation, however, lasted only about a month, and when Moon p came opposition Jupiter r (ruler of the 7th—marriage), they separated again.

In Stan's chart Mars p reached the conjunction of Ascendant r in December 1937, and then moved into the 1st house. This aspect did not occur in Sue's chart until 1943. Its influence on Stan's thoughts manifested in aggressive and daring actions, and the association with Mars men in his work made his conduct less restrained. He satisfied his drive for significance by doing work that most of his crew considered too hazardous for them to undertake.

Even in his attitude toward the opposite sex, there was one similarity between himself and his sister. Her boy friend before marriage was of wealthy parents, and she seemed unable to adjust herself to any less opulent male. And apparently only a girl having all the opportunities had any attraction for Stan.

The one girl he dated repeatedly in his new environment was the daughter of his boss, and his boss was a wealthy contractor. In 1940 he decided, at least temporarily, not to date her, because he felt financially unable to entertain her in his usual elaborate manner.

Throughout 1938 Sue lived with her mother and brother. Stan worked steadily and studied music. Venus p was conjunction Saturn r in the house of studies (3rd), Saturn also being ruler of the house of long journeys (9th).

In 1939, this aspect to Saturn was still operative, as was Jupiter p sextile Saturn p; and added to this, Mars p reached the friction (semi-square) aspect to Saturn r. Furthermore, the 7th and 12th houses were heavily accentuated through Sun p reaching the semi-square of Jupiter r, and Mercury p reaching the semi-sextile of Jupiter r.

She got along badly with her mother and Mars p semi-square Saturn r mapped the increasing friction between them—Saturn being co-ruler of the 10th (mother). Stan got along well with his mother, but the actions of his sister caused him much worry. Saturn is in the house of relatives (3rd).

Sue struck up a friendship with another girl about her own age. In March, while Moon p was sesqui-square Neptune r, brother and mother received a telephone call from Sue in another state: she told them she was returning to her old environment. She and her girl friend were driving (#26), and had given no intimation they were going on the trip until after they were on their way. They intended staying with Sue's relatives. The motive for the trip seems to have been about equally divided between her desire to get away from her mother and the possibility of reawakening the interest of the wealthy boy friend of earlier days. As might have been foretold, the trip was a disappointment. Sue quarreled violently with her relatives—mapped by Mars p semi-square Saturn r—and within two weeks she was on her way back to brother and mother.

After getting back to them she decided to go to work, and got a fairly good job (#52) with a jewelry company. Saturn is co-ruler of the house of business (10th), and Mars ruler of the house of work (6th).

Jupiter, Mars, Sun and Saturn were all aspected by progression in 1927 when Sue first began to have difficulty with her heart. It had given her some trouble since, especially when she smoked to excess. In 1939 these planets again were all aspected by progression. Probably the trip to her old environment had placed her under much strain and excitement. On October 20, 1939, when Moon p was inconjunct Jupiter r, she had a nearly fatal heart attack (#33).

Astrologically, in Stan's birth-chart the Sun, which rules the heart, is not so closely conjunction the Ascendant. By the first of April, 1940, his Ascendant p was conjunct Venus r, an aspect which she had in 1936. It was operative in his chart in 1939 when she took her abrupt trip. Furthermore, in his chart Midheaven p reached the conjunction with Moon r in September, 1940; an aspect she had in autumn, 1936.

He did not take a long journey, but Ascendant p conjunction Venus r which was operating at the time of her trip, and Midheaven p conjunction Moon r, which was within orb at the time she got her job with the jewelry company, mapped pronounced effects upon his life. The Venus aspect coincided with his dating the boss's daughter. In the autumn of 1939, the boss—who handled a number of construction gangs—placed Stan in charge of one group of from six to eight carpenters required to handle each job.

The promotion gave Stan additional prestige (#25) and a raise in salary (#14). It is unusual for a man of his age to be foreman over older men. He turned his attention to construction work as an opportunity to gain both the significance and financial advantages he craved. He decided, therefore, to give up vocal training that he might devote his energies to this type of work instead.

Despite the predisposition to accident (#16) mapped by Mars p semi-square Saturn r, he suffered no bodily harm. But his work, in addition to the responsibility of directing all the men working on one building, was highly dangerous. When an electrically driven skilsaw strikes a nail, it often causes a short circuit through the operator, who cannot let go of it. Sometimes men who use it are badly cut. Stan walked about on high places cutting off the ends of rafters with such a saw. It gave him a sense of significance to do the work nonchalantly which some of his crew considered too hazardous.

Under Ascendant p conjunction Venus r in her chart in 1935, Sue was given a car. Early in 1940, with the same aspect operating in Stan's life, he bought himself his first car—a good one. Mercury p was conjunction Venus r; Venus p was sextile Jupiter p; and Jupiter p was sextile Saturn p in the house of short journeys (3rd).

From the viewpoint of fortune-telling—which many astrologers find difficult to abandon—the divergency in the lives of Stan and Sue seem puzzling, even though progressed aspects at the time of some important event were not identical. But other charts handled by the Research Department indicate that the addition of one strong progressed aspect in the chart to the progressions operating in both, frequently maps an event to one person that is not present in the life of the other.

The planetary positions at birth map, but do not endow, the individual with the characteristics they indicate. Instead, the positions of the planets at any given time indicate the direction in which their energies are flowing and combining with other planetary energy currents. And the sum total of these energy streams and their convergences form an inner-plane energy pattern.

A pattern of steel is quite rigid and anything of size to pass through its apertures must rigidly conform in outline to them.

A similar pattern of rubber is more elastic. A glove of rather heavy rubber, for example, may be made to conform to average size hands which vary rather widely in detail, even if not elastic enough to permit entry of a large hand, or to fit a small hand. Nor will such a glove adapt itself to the foot or head. Likewise a pattern of swiftly moving water, as from the perforations in a nozzle, can be deflected within limits without splashing and spilling and spoiling the general lines.

Actual observation of people born with almost identical birth-charts indicates that astrological patterns have considerable elasticity. States of consciousness within the thought-cell organization exercise psychokinetic power to attract experiences corresponding to the planetary types; harmoniously or discordantly according to the formation of aspects; and relative to a phase of life mapped by the houses of the birth-chart. Not only do birthcharts of men but also those of animals — as determined by a study of such animal birth-charts in connection with events (#86)—map character traits. On the evolutionary level where the creature functions, and within the opportunities afforded by the physical environment, events are attracted clearly typical of these character traits.

The following appropriate quote is from "OCCULTISM APPLIED," by C. C. Zain:[4] "Instincts are common to other young animals, but are almost totally lacking in human infants. . . . Well defined emotions are not present in the infant at birth, but develop as the result of experience. . . . In other words, how a person acts later in life in the presence of a given situation depends upon the conditions surrounding similar situations that he was called upon to face earlier in life and particularly in childhood. . . . Experiments carried out on infants and children in the John Hopkins Hospital and the Harriet Lane Hospital, have given us much reliable information upon just how the manner in which a child reacts to a given circumstance is determined. It is determined, except the few unconditioned responses . . . and certain reflex actions of the muscles, by the experiences of the child after birth. That is, both the

[4]Brotherhood of Light Course XIV, Lesson 155.

mental attitude toward things and the skill with which actions are performed, all are acquired. They are not instinctive."

Yet while the mental attitude toward things and the skill of performance are acquired through the conditioning of physical environment, psychological experimenters recognize that the conditioning of two children by identical physical conditions may be widely different.

It is equally certain from charts and progressions already studied that a progressed aspect which occurs in a given chart when its owner is in one type of physical environment may not map the same event as the same aspect occurring when the physical environment is quite different. Often almost identical birth-charts and progressed aspects work out in quite different terms when the early conditioning physical environment has been dissimilar.

Some psychologists hold that with given hereditary tendencies, what a child becomes and does is determined solely through its conditioning by physical environment. And the fortune-teller type of astrologer claims everything that the child does and what happens to it throughout its whole life is laid out in inevitable rigid detail by birth-chart and progressed aspects.

Because a child lives in two worlds it must react to both. It is impossible for it to live so exclusively in the inner-plane world under the impact of planetary energies that it is not also influenced by physical conditions; and it is impossible for it to live so exclusively in the physical world that the energies mapped by progressed aspects do not influence it profoundly.

This means that the child is at all times being conditioned by two environments. The addition of planetary energy to specific groups of thought-cells empower them to attract events of their type into the life. The specific nature of the events attracted is not determined by planetary energies alone, nor by thought-cell activity alone, but also by the available physical events. In tropical lowland thought-cell activity will not, no matter what the progressed aspects are, attract death by freezing in a blizzard. Nor will such activity attract yellow fever to a native living in Greenland.

After birth the conditioning of the energies expressed by the groups of thought-cells and aspects become more specific, because physical environment affords specific things. It affords people who frown upon specific actions and who praise other specific actions, and whose emotional reactions have an effect upon the developing child. A 1st-house Sun in two charts provides an example. This may mean more volume to the Power urges in the life of one than in the life of the other, but the urge represented by a 1st-house Sun will be obvious in the trend of both lives.

It will certainly map a strong drive for significance. And the aspects of the Sun indicate channels mapped by some house and planet, through which significance will be sought. In the charts of the twins, Stan and Sue, Sun trine Uranus indicates that significance would be sought through some unusual behavior affecting the reputation. But the exact nature of this behavior is not so easy to determine. Significance indicated through a Mars-Saturn channel may be sought either in crimes of violence or in capturing dangerous criminals, and from the chart alone it is difficult to determine which. A child who gained significance in the slums as a tough guy, showing daring in petty crimes, usually continues in that direction. But a child who gains significance through accomplishing daring acts which are socially acceptable may end by being a G-man.

Definite strong urges in the birth-chart attract the person to corresponding behavior. Both the sex urge and the individualistic urge, unspecific in detail, express characteristically through a variety of specific events readily afforded by most environments. But expressing the drive for significance (Sun) through individualistic (Uranus) channels may take a variety of characteristic forms. Stan's behavior in finding significance through reckless spending and hazardous work is not less individualistic than Sue's behavior in finding significance through reckless companionship and athletics.

Perhaps, had the male twin been separated from his sister and older brothers, with a slightly weaker brother, he would have gained significance through athletic participation, if encouraged and praised.

CLASSIFICATION OF COMPARISONS

The same serial number will be used to denote identical or nearly identical progressed aspects and events attracted into the lives of the twins. These events may differ in one or more of five different ways, which will be designated by a single letter.

(S) *Specific Event*—Events may or not be of similar importance to the native, they may or not be similar in fortune or misfortune, they may or not belong to the same birth-chart house, and they may or not have the characteristics of the same planet; yet differ in Specific nature. For instance, a long journey, publishing, and a religious experience, similar in all other four classifications, differ as to Specific nature.

(I) *Importance of Event*—Judged by the importance to the native at the time and place where the event happens. Not considered are its importance to society, to any other evolutionary level, nor any situation other than that involving the individual.

(F) or (U) *Fortunate or Unfortunate*—Events similar in other respects may be fortunate for one individual and unfortunate for another.

(H) *House*—Two events related to different departments of life clearly mapped by different houses.

(P) *Planetary Characteristics*—Indicating two events clearly characteristic of two different planets.

When any of the five distinct differences are found between the events mapped by progressed aspects, the cause of the indicated difference may be any one or more of the seven factors denoted in The Church of Light Astro-Physical Interaction diagram:

(A) Character mapped by birth-chart.
(B) Special conditioning by reaction to events since birth.
(C) Volume and trend of particular progressed aspect.
(D) Volume and trend of all other progressed aspects.
(E) Resistance offered by physical environment.

(F) Effort at Control based on knowledge of progressed aspects.

(G) Evolutionary level.

These 12 letters—5 indicating the nature of the difference between events, and 7 indicating the apparent cause of that difference—provide classifications from which to compile easy reference tables.

With the classifications established upon which to base conclusions, the events in the lives of the fraternal twins will be studied for similarities and differences coincident with 15 sets of astrological configurations.

(1)[5] *Conclusion:* Regarding the attainment of significance, events in their lives differed specifically (S), but did not differ very much in importance (I). Sue attained significance through athletics, Stan through business enterprise. Sue's fortune in athletics is about equivalent to that gained by Stan in business. Stan's work and his early handling of live stock bear Mars characteristics as do Sue's athletic events. Causes of these differences appear to be in the birth-chart (A)—for sex is an essential part of the chart—and in the conditioning (B) by events since birth.

(2) Conclusion: The twins were unusually fond of their two brothers. When they left home, one in 1926, the other in 1927, the twins were affected emotionally and economically. But there is no indication that one twin was affected more or in a different manner by these events mapped by similar progressed aspects. Stan's added financial responsibility was paralleled by Sue's increased responsibility of caring for the home. Regarding the progressed aspects mapping emotions, finances and brethren, the attracted events appear to be similar in all five ways.

(3) Conclusion: While Sun p was semi-sextile Saturn r, Stan fell behind Sue in school. But at this time Sue had the addi-

[5]Number only in parenthesis indicates an astrological configuration appearing in both charts under consideration. For reference purposes it is designated in text as (See serial no. 23).

tional progressed aspect Ascendant p trine Uranus r. Difference in events here is not specific (S)—for both instances related to study—but in importance (I) and in the unfortunate (U) results for him. Events were mapped by the same house and had the same planetary characteristics. The difference stems from two things: Conditioning (B), and an additional progressed aspect (C) in one chart.

(4) Conclusion: In 1927 Sue began to have a health problem which Stan did not. The health difficulty, starting at this time and becoming serious late in 1939, appears to differ in all five ways: (S), (I), (U), (H), (P). It seems to be due to sex and other birth-chart (A) factors, to conditioning (B), and to additional progressed aspects (D). Sue also had Ascendant p semi-sextile Jupiter r.

(5) Conclusion: Both went to work in 1933 under Mars p square Moon r. The kind of work was specifically (S) different. The importance was similar, for Stan quit school and Sue formed associations which profoundly affected her life. These events were about as unfortunate for one as for the other, and the general planetary characteristics were similar. As school work and flirtations come under the rulership of the 5th house, Stan leaving school and Sue finding an affectional attachment come under the same house rulership. It seems, therefore, that the difference between the events was only specific, due to sex (A) and conditioning (B). Sue could not be expected to do a man's work, and her inclinations and conditioning toward both school and the opposite sex had been different than Stan's.

(6) Conclusion: The affectional attachments for both twins, mapped by the 1934 aspects, were not different specifically, in house rulership, or in general planetary characteristics, but differed in importance (I) and misfortune (U). Sue's affectional attachment and the good times she had were far more serious than were Stan's 7th and 5th house activities. There was an additional progressed aspect (C)—Midheaven p trine Venus r—in Stan's chart, and the environment (E) also had a bearing on the difference between the events. In considering the events in Sue's life, because she was of the opposite sex,

differences are caused through the birth-chart (A) and conditioning (B).

(7) Conclusion: The attitude toward the mother, and the influence of the mother on the life seems to be about the same in importance. This relationship was about as fortunate for both, was ruled by the same house, and had the same planetary characteristics. This attitude, however, was due almost entirely to difference in conditioning (B).

(8) Conclusion: Under progressed aspects to Uranus, at the time there were progressed aspects to the rulers of the 9th and 4th houses, Stan made a long journey and a radical change in residence, Thanksgiving 1935. The change coincided with Moon p trine Pluto r and square Neptune r; reenforced (#23) by Uranus m[6] conjunction Uranus r, and Sun m trine Pluto r and square Neptune r; and released (#91) by Jupiter t[7] inconjunct Pluto r, trine Neptune r and semi-sextile Moon p. Sue did not make the change until July 1936 when Moon p was inconjunct Mars p. For her it was timed (#23) by Sun m square Uranus r, and Moon m opposite Pluto r and inconjunct Neptune r; and released (#91) by Venus t conjunction Saturn p. It was practically the same journey for each. But as it was not taken at the same time, we can say it differed specifically (S) in timing. The girl and the mother could not take the trip until the boy had a job and could send for them. The difference in the time of the journey was due to environment (E).

(9) Conclusion: Aside from the journey itself, this move brought drastic changes into both lives. Under the stimulation mapped by Sun p sesqui-square Uranus r there was a breaking up of old conditions and attachments. Before the move, Sue's affectional life had been a failure, and even after moving to the camp her difficulties were increased due to lack of affectional attachments. Because he had always been shy, Stan's difficulties in adjusting to the rough construction workers were probably as great as Sue's problems. There does not appear to be dif-

[6]m, minor progressed position.

[7]t, transit progressed position.

ferences in any of the five categories. They both moved under Moon p parallel Saturn p, the trip timed by accessory energy mapped by Venus m and Sun m (ruler of the home—4th) trine Saturn p (ruler of journeys—3rd and 9th).

(10) Conclusion: Ascendant p was conjunction Venus r in March 1936 in Sue's chart and in April 1940 in Stan's chart. At the time the conjunction was operative the other progressed aspects were pronouncedly different in the two charts. But each twin under this aspect gained a certain social prominence; a romantic attachment and a first automobile. Considering that other differences in their lives were due to other progressed aspects, which seem adequately to account for them, the events coincident with Ascendant p conjunction Venus r were similar in all five ways.

(11) Conclusion: Apparently nothing in the birth-chart or progressed aspects accounts for Stan's taking pride in acquiring money and giving to others, or for Sue's finding an equal satisfaction in getting other people to give her money for her own pleasures. Money seemed to be equally important in both lives. The relation to money, however, is specifically (S) different. As the circumstances accounting for this different attitude are outstanding and known, it appears that the motivation should be attributed to difference in conditioning (B).

(12) Conclusion: When Sue married, May 7, 1937, she had an additional progressed aspect (C): Ascendant p conjunction Pluto r. This aspect, plus Venus p semi-square Mars p, and Moon p square Uranus r were operative not only at this time, but also on May 22 when she had her auto accident immediately after leaving her husband. These events were timed by Sun m on the cusp of the 7th house opposition Ascendant r, and Venus m in the 7th opposition Sun r; marriage, separation and accident all occurring while this accessory energy was being added. Moon m was sesqui-square Sun r on the date of marriage. Moon m was trine Sun r on the date of the accident. Both minor progressed aspects reenforced (#23) the major progressed Sun involved in the heavy aspects indicating these

events, which were released (#91) by Uranus t sextile Sun p, Jupiter t trine Mars r, and Venus t trine Sun r.

While Stan did not get married, he suffered discord as a result of his sister's marriage. He did not have an accident, nor did he take a long journey. These events in the lives of the twins were different specifically (S), different in misfortune (U), and in house rulership (H). Stan's economic adjustment was probably as important to him, but more fortunate, than the events which occurred to Sue; he began to acquire a sense of significance which he never before had. Yet there was turmoil, disturbances in the home and with associates, emotional upsets, and all the other general types of conditions characteristic of the progressed aspects in both charts. The difference between the events which occurred in the two lives seems to have been due to the addition of the heavy Pluto aspect (C) in Sue's chart, to her sex (A) and to her previous conditioning (B).

(13) Conclusion: In 1939 Sue took a trip and had much difficulty with relatives. Stan did not take a trip, but he did give up studies which he had pursued for several years, and he had difficulty over his sister's conduct and her trip. However, by this time he had an additional progressed aspect (C) within orb, Ascendant p conjunction Venus r. There were a number of changes in his work (Venus, ruler of 6th—work). The events in Stan's life were more fortunate (F), and they were specifically (S) different in the two lives. Events were about the same in importance, in house rulership and in planetary characteristics. The differences indicated seem to have been due chiefly to the additional progressed aspect (C) and to conditioning (B). Stan had become conditioned to consider his responsibilities more seriously.

(14) Conclusion: On October 20, 1939, when Sue had a nearly fatal heart attack, Sun p (ruling the heart) was semisquare Mercury r, and Moon p was inconjunct Jupiter r. The event was timed by Sun m (heart) conjunction Uranus r. Stan at this time had an additional progressed aspect (C)—Ascendant p conjunction Venus r—which may have mapped stimulation assisting him to retain his health. The cause of the dif-

ference was probably due to glandular sex differences (A), accentuated by conditioning (B). Sue craved excitement, toward which the closer trine of Sun p to her Uranus r undoubtedly mapped a stronger predisposition (due to sex) than the aspect mapped for Stan. And as in Conclusion 4, the heart attack coming to one and not the other is an event differing in all five ways: (S), (I), (U), (H), (P).

(15) Conclusion: Perhaps getting a good job with a jewelry company in the autumn of 1939 was not as important as becoming the boss of a crew of older men. But considering the economic level occupied by Sue, the jewelry job was as important to her as the foremanship job was to Stan. As a girl would not become a foreman of a carpenter crew at that time, there is probably not even a specific difference between these jobs. The difference was due to sex requirements and to the economic level occupied by each. Both advanced themselves where position and pay were concerned. Under the circumstances, these events are similar in all five ways. In Conclusion 5, concerning new jobs, sex was also a factor, but the effects of the jobs on the lives were specifically different.

TABLE OF CONCLUSIONS

Similarities and differences arising in the lives of these natural twins of opposite sex under 15 sets of astrological configurations:

Sets of Astrological Configurations Analyzed	15	100%
Events Specifically Different (S)	9	60%
Events Different in Importance (I)	4	27%
Events Different in Fortune (F or U)	5	33%
Events Different in House Rulership (H)	4	27%
Events Different in Planetary Characteristics (P)	2	13%
Difference Due to Birth-chart (A)	6	40%
Difference Due to Conditioning (B)	10	67%
Difference Due to Additional Progressed Aspect (C)	5	33%
Difference Due to All Other Progressed Aspects (D)	1	7%
Difference Due to Environment (E)	2	13%
Events Not Different in Any Category	4	27%

Events Different in Only One Way	4	27%
Events Different in Only Two Ways	4	27%
Events Different in Only Three Ways	1	7%
Events Different in Only Four Ways	0	0%
Events Different in Only Five Ways	2	13%

ASTROLOGICAL TWINS

Because case history studies of astrological twins proved so revealing in the research program, the following charts are presented. The same serial number will be used to denote identical or nearly identical progressed aspects and events attracted into their lives. Although the same serial number appears at the beginning of several paragraphs relating to the chief subject matter about aspects and events, different years may be involved.

Astrological twins are those born of different parents on the same day, or even when not born on the same day those who have nearly identical birth-charts. The following two cases fall into the later category, and the data was secured from each native personally. Douglas Fairbanks was born May 23, 1883, Denver, Colorado, 39N45 104W59, at 08:59:51 a.m., Local Mean Time. After his chart was erected, the research department located and received splendid cooperation from Doug's astrological twin. Fictitiously called Joe, he was born May 27, 1883, 42N00 75W00, at 8:54 a.m. Local Mean Time.

Although born four days later than Doug, Joe's chart is quite similar. Chief difference is that Joe has Moon in the seventh house, while it is in Doug's fifth. House cusps are only a few degrees at variance, although there are different signs on the cusps of the sixth and twelfth. Three planets are intercepted in Joe's chart which are free to express in Doug's, and Jupiter is intercepted in Doug's chart but not in Joe's. Aside from the Moon, all the other planets are in the same houses.

Events or conditions coinciding with 25 outstanding sets of astrological configurations were acquired concerning Doug's life; then Joe answered questions relative to these same periods of time. The results are presented below for easy reference, the same serial number given for each set.

(1) Although Joe moved to a new town earlier than the time of aspects now to be noted, there is no record of any specific event in Doug's life until he was 12 years old, and Mars p came to the sesqui-square of Uranus p and sextile Jupiter r. Mars rules the 10th and 5th houses in both charts. Psychokinetic energy added to the thought-cells relative to honor (10th) and to those relating to entertainment (5th) coincided with Doug getting a small part with a show when Steve Brodie, the man who jumped from Brooklyn bridge, appeared in it in Denver.

(1) The same two progressed aspects occurred in Joe's chart when he was 8 years old, and when in his chart—but not in Doug's—Jupiter p was still sextile Venus r (C) ruler of the home (4th). As this aspect had been within effective orb since birth—but not in Doug's chart—it probably accounts for the earlier move as well as the one to a different town at this time. Schools, as well as entertainment, are ruled by the 5th house, and the move made under these aspects placed him where he received regular schooling.

(1) While he did not go on the stage as an actor, even to do a small bit while one show was in town as did Doug, Joe did sing in the church choir, at school festivals and was elected as the only boy soprano from his school qualified to sing at the dedication of a high school, where he was surrounded by 500 girls. Though the specific circumstances were different, the honor (#25) Joe attained in entertainment when 8 was quite as outstanding as that attained by Doug at 12.

(1) Conclusion: The difference in specific events (S) seems to have been due both to special conditioning (B) and to difference in environment (E). The importance of the events was similar, they were about as fortunate for one as for the other, they belonged to the same birth-chart house, and they bore the same planetary characteristics.

(2) When Doug finished school, Sun p was semi-sextile Neptune r and square Uranus r and then moved up to opposition Moon r (C) in the house of stage and of stocks and bonds

(5th). Very restless, he shipped on a cattle boat to Europe, and tried working for a bond house. Mars p, ruler of profession (10th) and the cusp of the 5th (stage) sextile Jupiter p. This accessory energy mapped thought-cell activity enabling him to get on the stage, after some shifting about, where he continued.

(2) With the exception of Sun p opposition Moon r (C), these aspects occurred in Joe's chart when he was 13 and 14 years of age. Doug had a change in home life and occupational changes. Note that the 5th house rules both school and the theater. Important things could not be expected of a 13 or 14-year-old boy as could of one 17 or 18. Joe wrote: "There were many changes in my life around the age of 14. My stepmother passed away, and my mind was very upset at the time. I wanted to work and father wanted me to continue school. Finally, I went to work as clerk and stock boy in a large store." Yet after trying other occupations as did Fairbanks, Joe also finally landed in the 5th house. "Went back to school, however, later."

(2) Conclusion: The difference in specific events (S) undoubtedly was due to (B) previous conditioning, (E) difference in environment, particularly the immature age of Joe, and to (C) Sun p opposition Moon r in Doug's chart—a very powerful aspect. The difference in importance (I) was probably due to the same three factors. The fortune and misfortune at the time was similar. In the matter of house rulership (H), the aspect of Neptune seems to have worked out in terms of the 9th house (voyage) in Doug's chart, and more in terms of the 10th house (mother) in Joe's. We cannot be sure, however, that Joe did not have a religious (9th house) experience at this time, nor that Doug's mother (10th house) was unaffected. The restlessness and change characteristic of Uranus, and the seeking of Utopia, characteristic of Neptune, appear equally strong in the lives of both under these aspects.

(3) With the Moon in Sagittarius, the sign of sports, in the house of sports (5th), and trine Mars—instead of not aspecting Mars—there is no occasion to go beyond purely as-

trological indications to account for the tremendous and life-long interest of Doug in athletics (#75) and sports. Yet during his school years the dominant Mars did coincide with Joe's interest in such sports. He writes: "Was at that time very much interested in all sports; football, baseball, tennis and basketball."

(3) Conclusion: So far as events relating to athletics are concerned it seems that they were different specifically (S), different in importance (I), and different in the fortune associated with them (F), but were not different in house rulership (5th) nor in planetary characteristics (Mars). All three differences noted arose primarily from the character as mapped by the birth-chart (A), but were accentuated by special conditioning (B) when it became apparent to Fairbanks that athletic stunts could be made a pronounced asset in his profession as an actor on stage and screen.

Doug never did have Mars p square Moon r (C). Note that Mars in the 10th rules not only the job but also the mother, and is general ruler of army and navy. It is interesting to see what happened to Joe under this aspect that Doug did not have. With Moon in the 7th the environmental (E) factor of age prevented partnership which would have been expected had he been 10 years older at the time. Although the aspect did not map a partnership, Joe actively contacted the public (Moon in 7th).

When questioned about this, Joe wrote: "Around that time, 17 years, was when I had the idea of the cattle-boat trip [he did not take it, but Doug at about the same age did take such a trip]. I traveled around quite a bit that year working in different places. I joined the army in 1900, but my mother (father has married again) would not consent: so joined the National Guard. Tried the navy but wasn't heavy enough. In 1901 I went into the car shops and also to night school. Took three years' mechanical draughting [Mercury p was parallel Venus p during this time]. In the meantime had been promoted to timekeeper, shipping clerk, etc. The idea, I believe, was to teach me the several departments of the plant in order to make me some kind of an executive."

(4) Doug married in 1907. Two of the progressed aspects at the time—Venus p conjunction Pluto r, and Mercury p semi-sextile Mars p—Joe had in 1903. Joe reported that during 1902 and 1903 he was going steady with a girl, and the understanding was that they would be married. Joe writes: "I had lots of social life [Doug had become a matinee idol], and took vocal lessons, and was told by the professor that if I worked hard I would be a wonderful singer—I didn't work hard."

(4) Conclusion: The events which occurred differed specifically (S), differed in importance (I), and differed in the amount of fortune (F), but did not differ in the other two factors. In both lives the events related to the same houses (5th and 7th) and to the same planetary characteristics. Probably the most important cause of these differences was that Doug at this time had an additional powerful progressed aspect (C), although the conditioning factor (B) that Doug was an actor may have been almost as significant. As to difference in importance, Joe's 5th house activity (singing) lead nowhere, but with Doug it brought stage success. While Joe was engaged to be married, and Doug actually married, both engagement and marriage belong to the 7th house.

(5) In 1908 Doug's Venus p was conjunction Saturn r, and in New York he was advancing his theatrical career. This aspect occurred in Joe's chart in 1904 when "I moved to New York and immediately went to work on the street cars as conductor, advancing very rapidly to responsible [Saturn] positions in that line—inspector, dispatcher, clerk, etc."

(5) Conclusion: The 1908 event had to be ignored because it was unknown if Doug moved or not. The specific (S) events attracted were different, for both men were conditioned (B) to follow different vocations. Relative to the station in life each occupied, however, the importance, gain in fortune, house rulership (Venus being co-ruler of business and Saturn the ruler of the house of work—6th in both charts) and planetary characteristics (Saturn increase of responsibility) of the events were quite similar.

(6) In Doug's chart in 1907, and not in Joe's in 1902 or

1903, but also in 1907, there was the additional heavy aspect—Sun p sextile Mars r—ruler not only of the 10th, but also of the cusp of the house of children (5th). From his marriage to Beth Sully, Doug's only child was born (#21). Conception took place under Sun p sextile Mars r. Douglas Fairbanks Jr. was born Dec. 9, 1908, between 3:30 a.m. and 3.45 a.m., New York City. By this time the progressed aspect of Sun-Mars had moved beyond its orb, but Mars p was parallel Neptune r. No progressed aspect of the Moon was operative. The birth was indicated by the accessory energy (#23) of Jupiter m trine Mars r, Sun m conjunction Venus r, Mars m semi-square Sun r, and Venus m conjunction Mars r; and timed (#91) by Venus t opposition Mars p.

(6) As mentioned, Joe had an aspect, Sun p sextile Mars r, in 1906 and 1907, but not in 1902 and 1903, when he became engaged to a girl and while he was taking vocal lessons. This Sun-Mars aspect did not coincide in Joe's life with marriage or the birth of a child, but it did map his keeping steady company with the girl for several years. He also writes: "In 1907 I took civil service examination for policeman on the New York Aqueduct which was being built at the time. I passed high on the list, was appointed, and so employed until late 1909." The addition of one powerful progressed aspect (C) during a period when other progressed aspects (D) are operative in one of two similar charts, often maps an outstanding event or condition in the life of the person having the additional progressed aspect; the event or condition not at that time occurring in the life of the other.

(6) Conclusion: In Doug's chart progressed aspects occurred during 1907, only a portion of which were operative in Joe's chart during any one year. The volume of energy spread over four years in Joe's life seems to have mapped less psychokinetic power to bring important events to fruition. Under Sun p sextile Mars r the specific (S) events were different. This was due to difference in conditioning (B) and to the addition in one chart of a powerful progressed aspect (C). And the difference in importance (I) of the events may be attributed

to the same two factors. There was no great difference in the amount of fortune of the two sets of events attracted; they belonged to the same house, and bore the same planetary characteristics. Joe was in love (5th house) and engaged (7th house). As Doug was divorced after a time, it is debatable whether the one who failed to marry or the one who married, under these progressions, was more fortunate. The birth of a child and contact with the stage are 5th house events. Both Doug and his astrological twin, Joe, made progress in their careers under the astrological factors being considered.

(7) Under quite similar progressed aspects to the ruler of the 10th house, some already mentioned above, Doug went on the stage or in some manner furthered or hindered his career as an actor, and Joe acquired a position involving either mechanical ability or clerical training. Thus from the time each went to work until the end of life, the specific (S) events attracted under the stimulation of a progressed aspect affecting the career, such as a progressed aspect to the ruler of the 10th, were different. What is the cause of that difference?

(7) Even though many people having a powerfully active 5th house Moon have not gone on the stage, the Moon in this house of Doug's chart closely trine Midheaven undoubtedly mapped his stronger predisposition toward such work than is shown in Joe's chart. And it may be that this aspect (in Doug's birth-chart (A) and not in Joe's) mapped Doug's attraction at birth to a home where dramatic ability was pronouncedly encouraged. Other than the 5th house activity, however, the constants for a movie star (#1) are as outstanding in one chart as in the other.

(7) Granting (A) was significant, what about the conditioning factor? Doug's father was a lawyer and a deep Shakespearean student. Not only did Doug hear Shakespeare quoted at length, but as a boy he was taught to declaim long passages from Shakespearean plays. Joe said, "There was nothing in my early life that would have turned my attention to the stage. My father, grandfather and uncle were all blacksmiths, horseshoers; and father and uncle were general forge shop black-

smiths later. While going to school I was secretary in boys' clubs and had minor clerical positions." This is Joe's early environment.

(7) As the Moon is the planet of Mentality, its position in the Mercury decanate of Aquarius no doubt predisposed Joe more toward the clerical work which Mercury rules, while its position in the Mars decanate of Sagittarius in Doug's chart predisposed the movie actor quite as strongly toward action.

(7) Conclusion: While Doug's chart shows some natural aptitude for mechanical and clerical work, undoubtedly birth-chart (A) differences were partly responsible for the difference in vocations (S) followed by both these men. It would appear, however, that the conditioning (B) factors of environment were of equal power in diverting the professional energies into the two vocational channels followed. True, it would not have been impossible for Joe with such a background as he had to have become an actor, but it would have been far more difficult than with such early training as Doug had. Yet Joe's early clerical and mechanical contacts made it easy for him in adult life to find work in car shops, do clerical work, be a street car conductor, and work hard for a motor company. The vocations also differed in importance (I), and in house rulership (H); but if we consider the matter of fortune and misfortune aside from the importance factor, one had about as steady success as the other. Other than the house rulership, (Doug's activities belonging to the 5th house and Joe's to the 6th) both displayed the same Mars characteristics in their work.

(8) Having mentioned Doug's marriage and the birth of a child (Serial No. 6), it is interesting to note the difference in the two lives relative to the events ruled by each birth-chart house, occupied by the Moon. Doug, with Moon in the house of children, and erratic Uranus sole ruler of the 7th, had more wives and fewer children. Joe, with Moon in the house of marriage, and Scorpio and Sagittarius co-ruling his house of children, had more children and fewer wives. It may be that the movie colony, as some claim, is a specially favorable environment (E) in which to change marital partners. However,

the Moon in Sagittarius in the house of love affairs, trine Mars and opposition Mercury, would seem more responsible for the thought-cell activity which attracted Doug's two divorces and three wives. Also, as the Moon is essentially a domestic planet much given to home and family, its position in Joe's chart trining Sun and Saturn mapped the strong attraction of one wife and eight children.

(8) Conclusion: Where events affecting the family life in general are concerned they differed specifically (S), differed in importance (I) and in good fortune (F). As these differences in their lives seem to compensate relative to houses occupied by the Moon, perhaps the events did not differ in house rulership or in planetary characteristics. And while noting that conditioning (B) and environment (E) had some influence over these differences in their family lives, apparently the basic character mapped by the birth-chart (A) was chiefly responsible.

(9) Although Doug had poked fun previously at the "flickers," as he called them, in 1911 when Sun p came semi-sextile Pluto r and Saturn r, and Mars p came exactly parallel Neptune r (the movie planet) he changed his mind and went to Hollywood to get into the movies. For three years he had a rather hectic time, until in 1914 Sun p reached the semi-sextile (growth) aspect of Sun r, when he received his first favorable screen recognition.

(9) In Joe's chart Sun p (superiors) was semi-sextile Saturn r and Mars p was parallel Neptune p in 1909. However, in 1908, when Sun p was semi-sextile Pluto r, he had Venus p conjunction Sun r (C), which Doug did not have. By late 1909 this progressed Venus aspect, which Doug did not have in 1911, was out of orb of influence. During the first years of Doug's movie work, D. W. Griffith was highly displeased with his stunts and antics, and did all he could to discourage them. But Doug would not be dissuaded. Joe said, "In the latter part of 1909, through a little bull-headedness, if you want to call it that, I had an argument with the officer in charge of the station and quit the police and went to work for.........................." (a commercial firm with branches over the U.S.).

(9) Joe began traveling over ten states for this firm in September, 1910, under Jupiter p (the salesmanship planet) inconjunct Moon r (a progressed aspect Doug never had), Moon p conjunction Moon r, with Sun m square Uranus r in the house of travel (3rd) adding accessory energy (#23) and triggered (#91) by Jupiter t opposition Midheaven r inconjunct Neptune r and semi-sextile Uranus r. He writes of this period: "During 1909, 1910, 1911 and 1912 my life was rather hectic. Could secure jobs easily (#52), but seemed dissatisfied and would soon move on." These years of constant change were also coincident in Joe's chart with Mercury p (ruler of the house of short journeys—3rd) semi-square Mars r (ruler of jobs—10th). The aspect was in operative orb from 1909 until 1913, when it went out of orb and Joe settled down. This progressed aspect never did occur in Doug's chart. However, under other progressions later on in life, he experienced a period of roaming and restlessness, taking him even farther afield.

(9) Conclusion: Differences in the specific (S) nature of the events were due to occupational conditioning (B). Joe's incessant travel as a salesman was largely mapped by the additional progressed aspects (D) operative in his chart. In this matter of travel the events belonged to different houses (H). The other events in general concerning business belonged to the same house. Both men made changes, had altercations with the boss, and had a restless, hectic time. In the importance of these events, their fortune and misfortune, and their planetary characteristics (Mars—strife) were markedly similar.

(10) The other two progressed aspects which Joe had in 1910, when he became a salesman (#59), Doug had in 1914; Sun p semi-sextile Sun r (ruler of Ascendant), and Mars p conjunction Neptune r in the 10th house. 10th house progressed aspects are associated with business and the career. Having no contact with the movies, Joe could expect a possible promotion, fulfilled to some extent by his work for the commercial firm. But as Doug was already working in pictures, he could expect the event to affect his movie career. Knowing

the environment of each (E), different specific (S) events could be expected, but in each case characteristic of the 10th Mars and Neptune.

(10) Doug's first screen success (10th) was in "The Lamb," in which he gained recognition for athletic (Mars) stunts. Joe got a new job (10th), and during this period when Mars was conjunction Neptune, he drank (Mars) too much; apparently heavy drinking was not customary at other periods in his life. All people having a heavy progressed aspect to Mars do not take to drinking, but it is worth recording here that of the many persons who have taken to drink, in which the known dates of commencement of such excess were studied, there was a progressed aspect to Mars operating at the time in their charts.

(10) Conclusion: Due to vocational conditioning (B), specific (S) events affecting the careers can be expected to be different. On Joe's social and economic level, his progress with the commercial firm was as important as Doug's first screen success. The amount of fortune was similar, the same houses were involved—drinking implies convivial companions and entertainment (5th house)—and the planetary characteristics of the events were the same.

(11) Progressed aspects to Uranus coincide with new people entering the life. By 1916-17, having so thoroughly established himself in pictures, Doug was seen constantly in the company of Charlie Chaplin and Mary Pickford, under Mars p conjunction Neptune r; Mars p trine Uranus r; and Sun p semisextile Saturn p in the house of friends (11th). He reacted profoundly to this friendship, later becoming associated with them in picture production (#26). It was impossible for him to marry Mary Pickford at this time, as each was already married to someone else. But later this event took place. Their romance, widely publicized, drew the well-wishes of the nation.

(11) By 1917, in Doug's chart, not only these three aspects were operative, but Sun p was closely conjunction Jupiter r. Joe had the same four progressed aspects in 1913. Yet even though a difficulty of another kind arose, he had a certain advantage over Doug in whose chart, but not in Joe's, Mercury

p was semi-square Venus r (C), the planet of affection. Evidently the lady of Joe's choice was single, even though the Neptune aspect alone usually maps involved conditions; for on Oct. 6, 1913, after six weeks acquaintance (Uranus) he married her (#2). Moon p was then square Mercury r, but the event was more precisely timed by Sun m sextile Jupiter r, also reenforcing (#23) the major progressed aspect. The timing was triggered (#91) by Mars t inconjunct Moon r.

(11) Because a progressed Mars affliction mapped the difficulty related to his job, Joe's frank appraisal is pertinent: "In January 1914 I lost my job through bull-headedness again, and had a very hard time of it, working only a short time as a collector, at which I did very well. Went to work in June, 1914, in a canning factory and quickly advanced to foreman, but as this is only seasonal work it was rather hard going in the winter time." Heavy progressions involved Jupiter, Sun, Mars, Uranus and Neptune. The loss of job (#53) in January, 1914, was timed (#23) by Sun m applying conjunction Uranus r, ruler of other people (7th) and sudden change, and triggered (#91) by Jupiter t square Mars r (in 10th-job). When Sun m in June reached the trine of Saturn, ruler of the house of work (6th), Joe went to work at a different job.

(11) Under Sun p conjunction Jupiter r and semi-square Neptune p (planet of promotion) in 1917 and 1918, Doug did not get a job in a canning factory as Joe did; but, widely known to the public, Doug took a new job during World War I. He is credited with selling some three million dollars worth of Liberty Bonds for the government.

(11) While Doug's business life was not free from difficulty, the additional progressed aspect (C) affecting his emotional life (Venus), together with different environmental (E) conditions (he was already married) coincided with the trend of his difficulties. As the emotional condition surrounding Joe's marriage is not known, no valid comparison is possible. It is likely that Doug's companionship with Charlie Chaplin and also Mary Pickford (whom he afterwards married) was as important to him as marriage to Joe.

(*11*) *Conclusion:* Specific events (S) were different due to conditioning (B) and environment (E). The importance of the events was about the same to each. The amount of fortune and misfortune was similar, even though they arose from different departments of life. Friends and companionship were prominent in the lives of both. The 7th and 10th houses in both charts gained activity. The planetary characteristics of the events were similar.

(12) In 1918 Mercury p in Doug's chart came even closer to the semi-square of Venus r, an aspect Joe did not have until a number of years later (serial no. 13). But in both charts Sun p was semi-square Mars p, adding more thought-cell difficulties, and Uranus p (ruler of the 7th) was trine Neptune r. Events involving both wives (7th) took place. Understanding the difference in these events demands the consideration of Doug's chart where there was an additional progressed aspect (C) and a different environmental factor (E). It would not have been impossible for Fairbanks' wife to have had another child, as they had been married eleven years; but he was separated from her. It is to be assumed he was in love, because of active 5th house thought-cells mapped by the progressed aspect to Mars (ruler of the 5th cusp).

(12) From an environmental (E) point of view, it was much more likely that Joe, who had been married only the previous year, should gain a child. His son (the first of eight children) was born (#21) July 19, 1914. Moon p was square Jupiter r; but as the event was mapped by an aspect of the Sun to Mars, it was reenforced (#23) by Mars m sextile Sun r; and triggered (#91) by Mars t sesqui-square Mars r, and Uranus t conjunction Moon r.

(12) Although the legal action concerning Doug's wife took place earlier in reference to the other progressed aspects, Sun p was semi-square Neptune r (ruler of the 9th—court) when Joe married and Doug divorced. Doug's wife received her interlocutory decree on Dec. 1, 1918 (#10). As was the case when the first child was born to Joe (this affected his wife), Moon p was square Jupiter; in this instance, however, square

Jupiter p. Probably divorce was more important to Doug's prestige than the birth of a child was to Joe's. The heavy influence was the same: Sun p semi-square Mars p. But instead of Mars m aspecting the Sun, the divorce was timed (#23) by Mars m inconjunct Mars r, ruler of the house of prestige (10th). The event was triggered (#91) by Mars t sesquisquare Pluto r and Neptune r, and Venus t inconjunct Jupiter p.

(12) Conclusion: Once more the addition of a heavy progressed aspect (C) in Doug's chart and different environmental (E) conditions coincided with different specific (S) events. It is difficult to determine the relative importance (I) and fortune (F) of the events, because of an additional progressed aspect (C) in Doug's chart, but probably being divorced was more important to him and more fortunate than was the birth of a son to Joe. Both the 5th house (assuming Doug to have been in love) and the 7th were chiefly involved in the events affecting both lives, and the planetary characteristics (Mars) were similar.

(13) Although most of the nation applauded the Fairbanks-Pickford romance, there was gossip when Mary obtained her divorce from Owen Moore in Minden, Nevada. At this time and until after Mercury p was within the one degree orb of conjunction Venus p, the life of Fairbanks was not free from the annoyances mapped by a semi-square aspect. These semi-squares of Mercury p and Venus p to Venus r occurred in Joe's chart in 1920-21. But as at the time there was an additional powerful progressed aspect operative in Joe's chart (not present in Doug's)—Mercury p and Venus p square Uranus p—any astrologer would have expected Joe's events to be more severe.

(13) Fairbanks met Mary Pickford when Mars p was conjunction Neptune r and trine Uranus r (serial no. 11) in 1916-17, following which date Mercury p semi-square Venus mapped obstacles to marriage. Even though this progressed aspect was closer to perfect in 1920, progressed Mercury was moving so slowly that progressed Venus had nearly overtaken it (serial no. 14). On March 28, 1920, when Doug and Mary were mar-

ried (#2), Moon p was semi-square Venus p and conjunction Venus r. They first began to keep company under a Mars-Uranus aspect while Uranus p (ruler of the 7th) was still trine Neptune r. Mars m trine Neptune p and conjunction Uranus p (ruler of the house of marriage—7th) reenforced (#23) the major progressed aspect, and the marriage was timed by the trigger (#91) effect of Mars t trine Sun p, and Uranus t trine Jupiter r and square Saturn p.

(13) Mercury p and Venus p were in Gemini (ruling arms and hands) and progressed Sun was not quite out of orb of semi-square Mars p (accidents, and ruler of the 5th—children). By this time, Joe had another aspect long past in Doug's chart: Saturn p conjunction Sun r—Saturn ruling the house of illness (6th). Joe reported: "On Dec. 24, 1919, I had my hand smashed [mapped by aspect of Uranus and Mars] in the shop, losing half of one finger on the right hand [Gemini]." At this time Moon p was conjunction Mercury r, in Gemini. Mars m was retrograde and had not yet perfected its aerials to the conjunction of Uranus p and the square of Mercury p and Uranus p. The major progressed energy-releases were reenforced (#23) by Sun m square Saturn r, accentuating the major affliction of Saturn (ruler of house of illness—6th) conjunction Sun r. Further reenforcement (#23) effect was mapped by Mars m (accident) square Mercury r and square Moon p in Gemini (arms and hands). The accident was timed (#91) by Mars t square Jupiter p, and Neptune t opposition Moon r.

(13) Joe continued: "Feb. 8, 1920, a boy was born." No progressed aspect of the Moon was operative at the time. This event was timed (#23) by Sun m trine Jupiter r, co-ruler of the house of children (5th), and by Mars m, ruler of the house of children, trine Neptune r. The major and minor energies were released (#91) by Uranus t opposition Saturn r, Venus t opposition Jupiter p, and Jupiter t semi-sextile Jupiter p.

(13) Sun m square Saturn r, and Mars m conjunction Uranus p and square Mercury p and Venus p were within orb only a few weeks. Joe wrote of this period: "And I also had influenza [#32], but work was good and brought in good money,

I had a nice home [so did Doug] and everything was lovely until September, when a minor business slump came up and the shop was closed down [#53]. But I secured another temporary place at once and kept it until my own work started again in March, 1921 [#52]. However, all the children were sick [Saturn] with whooping cough, measles and diphtheria during the winter. They came out of it O.K., however." The change of employment in September, 1920, was not indicated by a progressed aspect of the Moon, but was timed (#23) by Sun m opposition Uranus p, and Sun m sextile Neptune p in the 10th. By 1921 Sun p was within orb of conjunction Jupiter p. When Joe resumed his previous work that year in March, Moon p was applying inconjunct Moon r, and the major progressed Sun-Jupiter conjunction was reenforced (#23) by Sun m square Jupiter r.

(13) Conclusion: The events mapped by Mercury p and Venus p semi-square Venus r were different specifically (S), and some of them were different in misfortune (U) and in house rulership (H). These differences probably were due to environment (E)—for Joe already having a wife could not marry—and to (C) the additional heavy aspects in his chart which seem largely coincident with both his accident and his other contacts with illness. Possibly the loss of part of a finger and the birth of a son were as important to Joe as was Doug's marriage to Mary, and aside from the events relating to the additional progressed aspects in Joe's chart, the events were similar in planetary characteristics. Home life and business were affected, some success and some friction relative to each in both lives.

(14) After the marriage, when Mercury p conjunction Venus p came within the one degree of perfect aspect, Doug and Mary acquired Pickfair, a beautiful home and estate, where they lived and entertained royally for the following ten years. 1920 to 1930 is called the Pickfair decade by the movie people. While it lasted Doug and Mary were pointed out as models of proprietv, and were idolized by the people of all nations.

(14) In Joe's chart, although Mercury p and Venus p did not make the semi-square of Venus r until 1920-21, Mercury came within one degree of conjunction of Venus p, ruler of the home (4th) in 1916. Even as Fairbanks settled down to a more sedate and stable life after some changing about, under Mercury p conjunction Venus p Joe did too. He moved to a new city to remain for the rest of his life. Mercury p and Venus p, moving at so nearly the same rate of speed, remained in conjunction over a number of years; in fact, in Doug's life throughout the whole Pickfair decade. Joe did not acquire so valuable an estate as Pickfair nor entertain so elaborately. In October 1922, under these progressed aspects, he purchased "five acres and a small house on the outskirts of town" which was his home for the rest of his life.

(14) In Joe's chart, but not in Doug's, at the time their respective estates were purchased, Sun p had reached the conjunction of Jupiter p. Jupiter sometimes maps good luck. Joe's purchase of a home later proved more lucky than the Fairbanks purchase. As will be related in due time (see serial nos. 17 and 18), an oil well was brought in on Joe's property in 1929. Moon p was sextile Pluto r at the time he bought this new and permanent home, but the purchase was timed (#23) by the two minor progressed aspects: Saturn m sextile Venus r, ruler of the home (4th), relating to its domestic value; and Sun m conjunction Neptune r and trine Uranus r, relating to the later unexpected (Uranus) discovery of oil (Neptune). The oil was not anticipated, nor was it discovered until Sun p reached the sextile of both Neptune r and Uranus r.

(14) Conclusion: Relative to Mercury p conjunction Venus p coinciding with both men settling down for a period of years, purchasing a home, and devoting their energies to domestic life and a career—on the social and economic level each occupied—the events attracted were strikingly similar in all five ways. The then unknown oil value of the property Joe bought may have been mapped by the additional powerful progressed aspect (C).

(15) Joe settled down in 1916, when Sun p was semi-square

Mars p (the mechanical planet), and got a job at a motor company doing heavy work. He held this job steadily 12 years, except for a temporary shutdown. Doug, having been conditioned to movie work, kept almost as steadily employed in pictures for 10 years. Mars rules Doug's 5th house, and it also rules the 10th house (position) of both charts.

(15) Conclusion: The difference relative to business and employment during some ten years seems to be specific (S) due to vocational conditioning (B). After considering the vocation and economic and social level each occupied, it is difficult to find much difference in any of the other four categories.

(16) At the end of the Pickfair decade in 1930, Mars p (planet of adventure and action) was conjunction drastic Pluto r and conservative Saturn r in Doug's house of friends (11th). With so much Mars energy activating the thought-cells mapped, Doug longed for action and adventure and began to loathe his life as leading exponent of screen respectability. The conventions (Saturn) weighed more and more heavily upon him. When he reached the point where he could no longer endure to travel the now well-worn rut, he left wife, home and friends to become a restless wanderer (#12) over the face of the earth. Mars p at this time was also semi-square Jupiter p, co-ruler of the house of long journeys (9th).

(16) In Joe's chart Mars p did not reach the conjunction of Saturn r until 1928 (serial no. 17), but when it did it mapped his quitting the job he had held so long—12 years with the motor company. But in 1926 (also in late 1925) Mars p was conjunction Pluto r and semi-square Jupiter p. Joe did not go to India to hunt tigers, yet Mars stimulated adventurous thoughts and those relating to friends. Even though differently conditioned by previous experience and current environment, Joe reported: "In December, 1925, I was arrested for possession of liquor. Came out all right on my good record. Work continued steadily, with good money coming in." Instead of a journey to a foreign land, as was the case with Doug, Joe's 9th house accentuation through the Jupiter aspect worked out in terms of law. Instead of hunting, Mars stimulation worked

out in terms of liquor and strife concerning it. Undoubtedly friends, signified by Pluto, were influential in both lives. Not the courts, but the law suit—or those who bring charges—is ruled by the 7th house. Joe's arrest was timed (#23) by a powerful minor progressed aspect: Sun m opposition Moon r.

(16) Conclusion: As far as they were specific (S) the events attracted under Mars p conjunction Pluto r and semi-square Mars p were different, apparently because of conditioning (B) and environment (E). Considering the social and economic level each occupied, probably there was no great difference in importance, in fortune, and none whatever relative to the houses involved and the planetary characteristics of these events.

(17) Joe had Mars p (ruler of jobs—10th) conjunction Saturn r (ruler of house of work—6th) in 1928-29, and as in Doug's life, work he had pursued for many years ended. The event making this change possible, however, was mapped for Joe by an additional heavy progressed aspect (C) Doug did not have until 1933-34. That event was the leasing of the 5 acres Joe had acquired in 1922—under Sun p conjunction Jupiter p—to an oil company which immediately started to work and brought in a small oil well. At the time this lease was made, Moon p was not making any aspect, but Jupiter m was inconjunct Uranus r and semi-sextile Neptune r (#23). The minor progressed aspect is significant because Sun p was sextile Uranus r and Neptune r. Thus major progressed aspects relating to the oil well were reenforced by minor progressed aspects. Leasing of the land on December 31 was timed by the reenforcing energy added to Mars and Saturn (planets in the heavier aspects having to do with employment), Saturn also the planet mapping things beneath the soil. Sun m was inconjunct Saturn r and Mars p, and Mars m was sesqui-square Mars r. The energies were triggered (#91) by Uranus t square Jupiter r in the house of hidden things (12th).

(17) Conclusion: While Mars p conjunction Saturn r mapped the condition affecting employment through different other events, they seem to be denoted by the different additional as-

pects (C) operative in the respective charts. As regards employment (Mars) and labor (Saturn) in these charts, there seems to be no difference in the events attracted in any of the five categories. Both quit work they had followed for many years to engage in activities totally different from those experienced before.

(18) In Doug's chart in 1933 and 1934, and in Joe's chart in 1929 and 1930, Sun p was sextile Uranus r and Neptune r, and sextile Uranus p and Neptune p. Doug had one powerful progressed aspect (C)—Venus p sextile Venus r—which Joe did not have. In 1916 Joe did not marry, but experienced other things parallel to those in Doug's life under similar aspects. 1920 Venus progressions mapped Doug's marriage, and in 1933 the additional progressed Venus aspect mapped a new romance for him, later resulting in marriage. Doug, the world wanderer, climbed rather high in British society. He met and fell in love with beautiful Lady Ashley. His former friends disapproved of his British aspirations and were scandalized by his companionship with the wife of Lord Ashley. Lord Ashley also objected and obtained a divorce.

Under a heavy Uranus progression new people enter the life significantly and usually there is a tendency to ignore conventions established by society. In addition to unexpected events, life undergoes so radical a change that things are never as they were before. One chapter of life closes and another opens.

(18) Other than the romance, accentuated in Doug's chart by the additional Venus progression, and Joe's oil-well-real-estate events mapped by Mars p conjunction Saturn r, the events were quite similar. Sun p sextile Uranus r and Neptune r in both cases mapped fulfillment of common expectations for such thought-cell stimulation; the difference between the two events mapped by additional progressed aspects. For instance, Doug traveled (#12) abroad (Neptune ruling long journeys—9th), and Joe wrote: "I also took a long trip [#12] back home to see the folks."

(18) Doug and Joe reacted to these new associates in such a manner as to change the course of their lives; Doug with the

British peerage, and Joe with real estate and lodge people. The oil company soon brought in a small oil well, and Joe received six thousand dollars (#14). Both rose somewhat socially, according to their own estimation; Doug through constant companionship with nobility, and Joe through ceasing 12 years hard labor, becoming active in lodge work, and becoming production manager of the oil company at a good salary (#52). Neither life was ever again as it had been before.

(18) Both embarked on unconventional interests, although in very different ways. Uranus in both charts is in the house of short journeys and studies (3rd), and rules the cusp of th 7th. Most people consider astrology an unconventional subject, and at this time Joe started the intensive study of astrology as a hobby, continuing until 1933. Even though both men took trips, the progressed aspect to Uranus mapped unconventional study for Joe, but for Doug, who also had a Venus progression, it mapped 7th house (Uranus) activity. Yet the 7th house was not inactive in Joe's life. In late 1929, he joined a fraternal organization recognized everywhere for its prestige, and advanced many degrees. He asserts this had a marked influence on his life, and he gained a companion: "I also became friendly with a man who in a way influenced me to do things sometimes against my better judgment . . . I was connected with him in various ways until December, 1936."

(18) Conclusion: Under Sun p sextile Uranus r and Neptune r most of the events concerning the two lives were essentially the same in all five ways. Those different specific (S) events seem to be indicated clearly by the additional progressed aspects (C) which at the time differed in the two charts.

(19) Wide, adverse notoriety Doug received at the time Lord Ashley divorced Lady Ashley coincided with Mars p (ruler of the 10th—reputation) making the conjunction (prominence) aspect with Sun r (serial no. 23), ruler of his personality (Ascendant). Joe did not have this aspect at the time the Uranus and Neptune progressions were operative; therefore his thinking did not attract such criticism. However, the aspect was in force in December, 1936, when Joe parted

with the man he had befriended under the 1929-30 Neptune and Uranus progressions. When Doug's prestige suffered so heavily and he lost popularity, Mars p was conjunction Sun r, and Mars p was also semi-square Midheaven r (serial no. 21). Although later planning to become a picture producer, his career slumped into an eclipse from which he never recovered, the slump mapped by these progressions.

(19) Conclusion: Mars p conjunction Sun r mapped difficulty with associates (11th) and loss of prestige (10th). The addition of Mars p semi-square Midheaven r (C), together with the environmental (E) factor that Doug was known throughout the world, gave wider publicity to the criticism which befell him. However, the sudden breaking of all relations with Joe by his business associate of six years, and the effect upon his business future, were as important and as unfortunate for him as was the wide criticism to Fairbanks.

(20) On March 6, 1936, while Venus p was conjunction Jupiter r, Doug married (#2) Lady Ashley. Uranus p (ruler of the 7th) was still trine Neptune p. No Moon progressions were operative, but for several months (#23) Jupiter m had been opposition Venus r, then squaring Uranus, the planet mapping their earlier acquaintance, and ruler of the marriage house (7th). Trigger effect (#91) was mapped by Venus t inconjunct Jupiter p, and Uranus t semi-sextile Sun r. After some adventuring Doug brought his wife back to Santa Monica, California, to a fine house by the sea where they expected to spend the rest of their lives. Venus is ruler of the home (4th).

(20) Joe had Venus p conjunction Jupiter r in 1932. Concerning this period, he wrote: "Domestic life was very good at the time. We were happy, only marred by the very serious illness of my wife in May, 1932. Children all well. Visited by relatives we had not seen in some time." Significant of the wife's illness, Saturn p (ruler of the house of sickness—6th) was conjunction Sun r; and Sun m, on the cusp of the home (4th), was sesqui-square Uranus r (co-ruler of wife—7th).

(20) Conclusion: Joe's wife's illness was perhaps mapped by the additional progressed aspect from Saturn. In so far as

marriage was concerned (S), the events were necessarily different for the environmental reason that Joe already had a wife. Nevertheless, events relative to his wife were prominent in his life, and seemed quite as important to him, even though their unfortunate nature was signified by Saturn (illness). Old associates, in Joe's case relatives (3rd) and in Doug's case neighbors (3rd), came back into both lives. House rulership and planetary characteristics of the events—other than the illness—were the same.

(21) Before the end of 1932, however, the same aspect—Mars p semi-square Midheaven r—which mapped the permanent obscurity of Doug's career (serial no. 19), occurred in Joe's chart. Astrologically, the effect for Joe at this time was far more severe than for Doug, because Saturn p for several years had been conjunction Joe's Sun r, an aspect Doug had when he was 17 and 18 years old and just starting on his own. While domestic conditions on the whole were more fortunate than usual, business matters started into a decline from which they did not recover.

(21) Joe described this period: "We were compelled to sell what was left of the business venture at a $500 loss [#15], also lost our automobile." The next year, 1933, he went flat, from the standpoint of business and employment. By this time Venus p had passed the conjunction of Jupiter r; Saturn p was only 26 minutes past the perfect conjunction with Sun r; and Mars p was semi-square Midheaven r. Joe said: "I did not work at all during early 1933. Went on C. W. A. in December, 1933. Health was excellent." Under these discouraging work conditions Moon p was making no aspect, but Sun m (ruler of the personality—1st) was opposition the planet of sudden change, Uranus. The change and its precise nature was timed (#23) by Mars m (ruler of the 10th—position) sextile Saturn r (ruler of the work—6th) and sextile Pluto r (ruler of group activity). These minor progressed aspects reenforced the major Saturn progression.

(21) Conclusion: Mars p semi-square Midheaven r mapped the permanent business eclipse of both men. Both made spor-

adic attempts at recovery, but were equally unsuccessful in staging a comeback. Apparently, from an astrological standpoint, this condition is the same in both lives in each of the five possible ways for events to differ.

(22) Although Joe experienced accidents until late years his health, like Doug's, was exceptionally good. He reported: "No sickness in my life with the exception of when I was 31, at which time I had a very severe attack of pneumonia, but came out of it O.K." Progressed Mars, plus progressed afflictions to negative planets, plus progressions involving planets in Gemini or Mercury are the constants for pneumonia (#18). As Doug's health condition in 1918 is unknown, no comparison can be made, but in 1914 when Joe suffered pneumonia Mars p was conjunction Neptune p; Sun p was semi-square both Neptune p and Mars p; and Venus p in Gemini was semi-square Mars r.

(22) Conclusion: While data on accidents in Doug's life is not available, both he and Joe boasted of habitual good health. Both were singularly free from illness of any kind up to the period when progressed aspects coincided with Doug's death. After that time Joe's health was not so good. In the absence of more precise data no difference was tabulated between them in the matter of health until after the time when, under Mars p semi-square Midheaven r, the fortunes of both underwent great change.

(23) With no previous indication that all was not as it should be Doug on December 11, after attending a football game, complained of severe pains about his chest and arms, and was ordered to bed by his physicians who said he was suffering from a slight heart attack. He went to sleep as usual, and at 00:45 a.m. on December 12, 1939, died without waking. Joe continued to live in fair health. At the time of Doug's sudden death (#3) Mars p was conjunction Saturn p; Mercury p was semi-square Saturn r and Pluto r; and Moon p was sesquisquare Saturn r and Pluto r. Reenforcing energy (#23) was mapped by Neptune m conjunction Pluto r and Saturn r.

(23) Doug had lived strenuously and until 1934 extraordi-

nary athletic stunts were a part of his daily routine. He had built up muscles and a heart capable of standing terrific exertion, but incapable of adapting themselves to a life of physical relaxation and idleness. When, through his attachment to Lady Ashley, he let up on his physical training to live a life of greater ease, the deterioration common to men who retire from strenuous life set in. His life underwent a drastic change noticeable in his physical appearance while Mars p was conjunction Sun r. He took on considerable flesh. He did not, however, complain of any tendency toward illness.

(23) In 1935, the year after the same progressed aspect was perfect in Doug's chart, Mars p made the conjunction of Sun r in Joe's chart. Although in the long run the effect mapped by the aspect was not so disastrous, the manifestation upon Joe's health was more obvious. The two planets were conjunction in Gemini, ruling the respiratory region. Joe wrote: "I have suffered for about five years [since this aspect] with what the doctors call chronic bronchitis. It seems as if my throat is cramped at times, cutting off oxygen to the heart, causing fainting spells. However, it has cleared up some since December, 1939, when I expelled a tapeworm."

(23) Again it should be noted that Mars p was conjunction Sun r and Saturn p; and Mercury p was semi-square Saturn r and Pluto r, in December, 1936, when Joe parted company with real estate work and his previous co-worker (see serial no. 19 for effect in Doug's career and associations). In Joe's chart Moon p was square Mars r in the career house (10th). The event occurring then seems to have severed the last thread of his relationship with private enterprise. It was timed (#23) by Mars m—one of the planets involved in the heavy progressed aspects—conjunction Saturn r (ruler of the house of work—6th), another of the planets involved in the major progressions.

(23) Conclusion: Without some knowledge of what was happening to Doug's heart in 1934 and after, it is hard to appraise the health influence mapped by Mars p conjunction Sun r. If Joe was told he had the worst of it, he could point out

that he lived longer than Doug and gained in health later. If it is assumed Doug had the worst of it, there is no actual proof he suffered at all in health at the time—other than gaining weight. Perhaps health difficulties under this aspect differ in the matter of misfortune (U) but not in the other four categories, and are chiefly due to past conditioning (B) by events. Their relative importance can hardly be judged, and if Doug was affected in health these events were of the same specific nature, belonging to the same houses, and had the same planetary characteristics.

(24) Although the heavy progressions operating in Doug's chart at death were operating in Joe's chart in 1936, Sun p did not reach the perfect square of Mars r until 1937, at which time Mercury p (then past the orb of semi-square Saturn r and Pluto r) was closely conjunction Jupiter p. Joe wrote: "In 1937-38, while I worked for WPA as foreman, clerk and diferent rated jobs above common laborer, my finances were very, very low. I was just making enough to live, and in 1937 the loan was foreclosed on my property, and it was only through the influence of some good friends [Mercury p conjunction Jupiter p] that I was able to buy back by paying as much as I had in the first place."

(24) Conclusion: Financial conditions in Doug's life just preceding his demise were favorable; and at that time he was preparing to produce movies. Perhaps business and work were less unfortunate (U) for him than for Joe, even allowing for the economical level of each. From a work standpoint this was expected as Saturn (ruler of the house of work—6th) was closely conjunction Sun r (C) in Joe's chart, but not in Doug's. In importance, specific type of events (relating to business and work), house rulership and planetary characteristics, these events were similar.

(25) Mercury p was semi-square Saturn r and Pluto r in Joe's chart in 1936 when he severed his contact with private employment (serial no. 23). But Mars p conjunction Saturn p (33 minutes from perfect in Doug's chart when he died) was operative in Joe's chart in 1936-37. Progressed Saturn was

but 38 minutes past perfecting its aerial on January 31, 1938. Also, on this date, Sun p was only 2 minutes past a perfect square with Mars r. These aspects map an accumulation of Mars energy in Joe's life on this date, which was at least as powerful as that in Doug's life on December 12, 1939. However, Joe's chart mapped one advantage: Mercury p sextile Neptune r and Uranus r.

(25) Joe wrote: "On January 31, 1938, I suffered an accident that could have cost me my life very easily. A 2900 pound steam-shovel bucket fell on me and broke 7 ribs in 9 places, 4 ribs on the right side and 3 on the left. I was laid up for 6 weeks, and also had to have nearly all my teeth [Saturn] taken out." At that time Moon p was inconjunct Venus p, but the reenforcing (#23) energy for the accident (#16) was mapped by Mars m conjunction Merucry r, Mars mapping the heaviest progression at the time. The event came under the trigger (#91) mapped by Mars t sextile Saturn r.

(25) Conclusion: Healthwise these aspects mapped an event which was more important (I) and more unfortunate (U) to Fairbanks. It can be argued that the more favorable progressed aspect in Joe's chart—Mercury p in good aspect (sextile) to both Uranus r and Neptune r, instead of Mercury p semi-square Saturn r and Pluto r—denoted his escape from death. Possibly if Doug had been conditioned by events since birth (B) as had Joe, he would have lived; during his life athletic work placed a tremendous strain upon his heart, offering the environment in which thought-cell activity mapped by progressed Mars aspects could operate more disastrously. On December 11 Doug attended a football game. He complained of severe pains about his chest and arms. It was Joe's ribs that were broken. In each case, the same general region of the body was affected, both suffered severe pain, and while a heart attack is not exactly the same as an attack from a steam-shovel bucket, astrologically, these two events may be considered specific, ruled by the same houses, and having the same planetary characteristics.

TABLE OF CONCLUSIONS

ASTROLOGICAL TWINS OF THE SAME SEX

Sets of Astrological Configurations Analyzed	25	100%
Events Specifically Different (S)	17	68%
Events Different in Importance (I)	7	28%
Events Different in Fortune (F or U)	8	32%
Events Different in House Rulership (H)	4	16%
Events Different in Planetary Characteristics (P)	0	0%
Difference Due to Birth-chart (A)	3	12%
Difference Due to Conditioning (B)	15	60%
Difference Due to Additional Progressed Aspect (C)	8	32%
Difference Due to All Other Progressions (D)	0	0%
Difference Due to Environment (E)	8	32%
Events Not Different in Any Category	5	20%
Events Different in Only One Way	10	40%
Events Different in Only Two Ways	3	12%
Events Different in Only Three Ways	7	28%
Events Different in Only Four Ways	0	0%
Events Different in All Five Ways	0	0%

Although these Case History studies have considered only similarities and differences in events relative to 40 sets of astrological configurations, many other studies bear out the inference to be drawn from these two published tables. They show the power of environment to condition (B) an individual so that the events attracted become different than had he previously been conditioned by another type of environment.

An appropriate quote appeared in TIME, March 11, 1940: "Few years ago a psychologist named Harold Manville Skeels, a professor at University of Iowa, was assigned by the State to advise the State orphanage. He found that the orphanage was sending babies (mostly bastards) born of feeble-minded parents to highly intelligent families for adoption. Horrified, Dr. Skeels hurried forth to see how much damage had been done. He gave the adopted children intelligence tests. To his surprise, their average I.Q. was 115, well above normal (100); Not one was dull."

As explained in "Cosmic Alchemy", by C. C. Zain[8], in the chapter entitled *Heredity and Environment,* it is only when a defective gene transmitting a definite characteristic is paired with another defective gene transmitting the same characteristic that the offspring is defective in this respect. A father with some defects and a mother with some defects may produce offspring superior in all ways.

Example of heredity combinations. Open circles represent normal genes; shaded circles, defective genes.

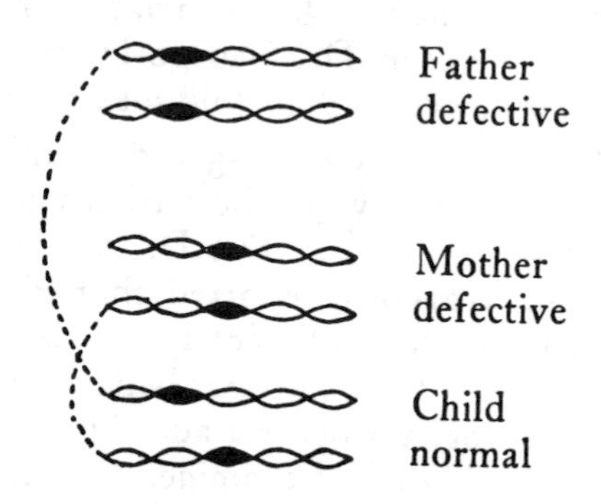

This diagram refers to children superior to both parents: Feeblemindedness, indolence, incorrigibility, irritability and other deficiencies often arise in different people from genes that are not of the same pair. Two people may be feebleminded, but not feebleminded in exactly the same way. Two people may be deaf, or blind, and the deafness, or blindness be a slightly different kind, arising not from the same, but from adjoining pairs of genes. Of course, in people who are closely related, a defect affecting a given part usually arises from the same pair of genes, but with people not closely related it may well arise from genes lying in quite different pairs.

The defective gene for feeblemindedness may come from one pair that are recessive in one individual, and from another pair of genes that are recessive in another individual. These two, when they unite, may then contribute one dominant gene

[8]Brotherhood of Light Course XVII, Lesson 167.

to each defective pair of the other, so that both pairs of genes in the offspring have one normal gene. Thus from two feeble-minded parents it is possible to get children who are bright and intelligent, superior to both parents.

Continuing with TIME quotation: "Dr. Skeel's discovery was one of a series made more or less by chance during the last 15 years by the University of Iowa's Child Welfare Research Station. The Station found that when children attended a nursery school or were transferred from a bleak orphanage to good homes, their I.Q.'s invariably improved. Concluded the Station's director, Dr. George Dinsmore Stoddard: with good upbringing even a dull child may become bright."

Not all the psychologists agreed to this and, "Last week, with blood in their eyes, the nation's leading psychologists gathered in St. Louis . . . But Dr. Stoddard had found supporters. The yearbook reported that identical twins reared in separate homes had different I.Q.'s. Southern Negroes who moved to Harlem (and thus got better schooling) raised their I.Q.'s. Psychologist Robert Ladd Thorndike (son of famed Edward Thorndike) had examined the records of some 1,000 children in three famed progressive schools (Horace Mann, Lincoln, Ethical Culture) found that in two schools children's I.Q.'s were static, but in the third (unidentified) there was an average I.Q. gain of more than six points.

"As the convention opened, bald, husky Dr. Stoddard rose to report. He twinkled: 'Only a confirmed Pollyanna would say that, as a committee, we have performed our task in a spirit of loving kindness . . .' Listeners looked at each other, wondered when the fireworks would start. They never did start. Long before the end it had dawned on the delegates that there would be no debate, because no one cared to get up and contend that nurture had nothing to do with intelligence. Said the final speaker, University of Chicago's Sociologist Ernest Watson Burgess: '(The) consensus (is) that intelligence, at least as measured by the I.Q., is not a constant and that it is a resultant both of heredity and environmental factors!"

From these two case histories, as well as many others studied, it seems equally certain that the events, which, to persons with identical or nearly identical birth-charts, coincide with the same progressed aspects, are determined not by progressed aspects alone, but are a resultant of thought-cell activity mapped by these progressed aspects (D), special conditioning since birth (B), and the environment (E) at the time.

As the two great weaknesses of scientific astrology today are the popular misconceptions of (1) what an individual can do in relation to what is shown astrologically, and (2) just what astrology claims to do and does not claim to do, we published the following brief conclusions.

Man has little or no control over the following two factors: his character at birth as mapped by the birth-chart, and the progressed aspects which accentuate and release the energies thus mapped at times which can be predetermined. But he has much control over two other factors of vast importance: the conditioning of his character after birth and the physical environment which facilitates thought-cell activity, mapped by a given progressed aspect, bringing the desired event and resisting the type of events not desired.

Here in three brief paragraphs is stated what astrology does, and what it does not do:

1. The birth-chart DOES NOT map SPECIFIC EVENTS, but it DOES map PREDISPOSITIONS toward the types of conditions and events which will effect each of the 12 departments of the individual's life.

2. Progressed aspects DO NOT map SPECIFIC EVENTS. But they DO map ENERGY-RELEASES of definite types during limited periods of time which can be predetermined. These energies, during such periods of release expend themselves bringing into the life conditions and events of their particular type, harmony or discord, and relating to one or more of the departments of life associated with them revealed by the houses which the planets mapping the energy-releases rule.

3. With sufficient knowledge of Conditioning (how the individual has reacted to events since birth) and the present physical environment (offering facilities for one specific event

and resisting others) often it is possible to predict events, not because specific events are inevitable. But because in the given physical environment the probabilities may be much more than 100 to 1 that the character as modified since birth, receiving energy-releases mapped by definite progressed aspects, will attract these SPECIFIC EVENTS rather than others.

IX
GLOSSARY

The following section of astrological terms embraces 506 words: some used in this book; in the Hermetic System of Astrology and Church of Light Astrological Research Department Bulletins; in common usage; and a few becoming obsolete in modern astrological parlance. If detailed definitions were given, they would fill volumes; therefore, the aim has been to keep these definitions as brief and non-technical as possible. The following words are explored at length, with abundant description and example, in the twenty-three books comprising the 21 Brotherhood of Light Courses. See forthcoming Volume XXII — *Index to the 210 Brotherhood of Light Lessons.*

ABBREVIATIONS

Asc.: Ascendant
C. D.: Calendar Date
C. T.: Calendar Time
C. T. I.: Calendar Time Interval
EGMT: Equivalent Greenwich Mean Time
EGMTI: Equivalent Greenwich Mean Time Interval
GMT.: Greenwich Mean Time
I: Interval
I. C.: Nadir
Lat.: Latitude
L. C.: Lunar Constant
L. D.: Limiting Date
LMT: Local Mean Time
LMTI: Local Mean Time Interval
Long.: Longitude
Log: Logarithm
m: Minor Progressed Position
Map.D.: Major Progression Date
Map.T.: Major Progression Time
M. C.: Midheaven
M. C. C.: Midheaven Constant
M. E. D.: Minor Ephemeris Date
Mip. D.: Minor Progression Date
Mip. T.: Minor Progression Time
p: Major Progressed Position
r: Birth-chart Position
R: Retrograde
S. T.: Sidereal Time
S. T.: Standard Time
t: Transit Progressed Position

Acceleration: Increase of planet's apparent motion relative to the earth. Synonym: Increment.

Accidental Dignity: Volume of energy expressed by a definite subconscious urge as mapped by a planet and determined by the particular house position of that planet.

Accidental Station: House position of a planet.

Adverse: See Discordant.

Aerial: A line stretching across the astral body, picking up and radiating types of energy mapped by planets involved in an aspect.

Affliction: See Discordant.

Agitation Compound: Conditioned association (mapped by the sesquisquare aspect) of two subconscious mental urges (mapped by the two planets involved in the aspect).

Air Movement Chart: Used in weather predicting; chart erected each time Mercury, either by direct or retrograde motion, enters a different zodiacal sign.

Air Signs, Triplicity of: Gemini, Libra, Aquarius.

Algebraically: Use of algebraic processes.

Algebraic Sum: Aggregate of two or more numbers or quantities taken with regard to their signs, as plus or minus, according to rules of addition in algebra; thus the **algebraic sum** of -2, 8 and -1 is 5; opposed to **arithmetical sum,** which supposes only positive signs.

Amplifier: Term used for Midheaven where all astral energies reaching the point it maps in the astral body are widely broadcast.

Angles: Points dividing the birth-chart into four quadrants: Ascendant (1st house cusp), Midheaven (10th house cusp), Descendant (7th house cusp) and Nadir (4th house cusp).

Angular Houses: First, Fourth, Seventh, Tenth. Houses of strongest volume.

Antidote: Remedy to counteract the effects of poison. See Mental Antidote.

Apex: Top of birth-chart; Midheaven.

Applying Aspect: Planets moving toward the completion of a perfect aspect.

Aquarian Age: Period during which the vernal equinox apparently travels backward through the constellation of Aquarius. Commenced January 19, 1881.

Aquarius: Eleventh sign of zodiac. "I know" motivation. Positive thought-cell activity, altruism; negative, argumentation.

Aries: First sign of zodiac. "I am" motivation. Positive thought-cell activity, leadership; negative, officiousness.

Ascendant: Point where planets rise on eastern horizon, mapped by first house cusp, located on extreme left of birth-chart.

Ascending Planet: One stationed in any house on eastern half of birth-chart.

Aspect: Angle at which astral vibrations from two planets meet. In the Hermetic System ten such angles where energies meet coincide with a definite influence. See Standardized Symbols of Aspects.

Association Houses, Trinity of: Third, Seventh, Eleventh.

Astral: Pertaining to the stars.

Astral Body, or Form: Mold of the physical body composed of astral substance organized by states of consciousness, functioning on a four-dimensional level where the total experiences of the soul are recorded. Synonym: Inner-plane body.

Astral Cells: Structural and functional units within the mass of psychoplasm forming an astral organism.

Astral Plane: Area of fourth-dimensional activity. Non-physical; first interior level relative to material plane.

Astrodynes: Units of astrological power.

Astrologer: One who erects and judges horoscopes, pointing out the necessary precautionary actions.

Astrological: Adjective referring to astrology.

Astrological Age: Period during which vernal equinox apparently moves backward through one constellation. Approximately 2,156 years.

Astrological Signature: A key, corresponding to one astrological factor determined by the type of energy radiated, associated with a particular manifestation.

Astrological Signatures: One of the seven branches of Astrology based upon painstaking observation of existing parallels between heavenly positions and worldly affairs. Its application indicates the relation between astrological factors and spiritual ideas, as well as between astrological factors and religious or other periodical observances on earth. Its function is to determine the kind of chart and type of progressed aspects used in natal astrology, horary astrology, mundane astrology, stellar diagnosis and weather predicting.

Astrological Vibrations: Frequencies radiated by the planets. Second most important group of imponderable forces to which reaction takes place. See Imponderable Forces.

Astrology: The science and art of interpreting basic key-tone frequencies, radiated by signs and planets existing at any given moment in time-space relationship, equaling basic subconscious urges (natal), ideas and questions (horary), general mass mind stimulation (mundane), moral ideas and precepts (spiritual), and electro-magnetic phenomena (weather predicting) as related to human beings and other entities, individual and collective, and to their minds, emotions and environments—all of which respond to the law of resonance, a basic principle in nature. See Branches of Astrology.

Autumnal Equinox: Point where the sun following its apparent path, the ecliptic, crosses the celestial equator in the Fall.

Barren Signs: Aries, Leo, Capricorn.

Benefic: See Harmonious.

Best Aspect: One having most harmodynes in a chart.

Best House: One containing most harmodynes in a chart.

Best Sign: One having most harmodynes in a chart.

Birth-chart, adj.: Referring to natal astrological factors: planets, signs, aspects, houses.

Birth-chart, noun: Erected for date, time of day and place on earth of a birth, accurately mapping outstanding character factors as they are at that moment.

Birth-chart Constant: A permanent value in the natal chart.

Branches of Astrology: (1) Astrological Signatures. (2) Natal Astrology. (3) Horary Astrology. (4) Stellar Diagnosis. (5) Mundane Astrology. (6) Spiritual Astrology. (7) Weather Predicting.

Business Planets: Saturn, Jupiter.

Cadent Houses: Third, Fifth, Ninth, Twelfth. Houses of weakest volume.

Calendar Constant: Synonymous with Limiting Date.

Calendar Date: Specific day, usually referring to an event.

Calendar Time: A number of years, months and/or days.

Calendar Time Interval: Time which has elapsed or will elapse between two dates.

Calendar Year: Usually refers to time interval between the Limiting Dates of two consecutive years.

Cancer: Fourth sign of zodiac. "I feel" motivation. Positive thought-cell activity, tenacity; negative, touchiness.

Capricorn: Tenth sign of zodiac. "I use" motivation. Positive thought-cell activity, diplomacy; negative, deceitfulness.

Cardinal Signs: (Term not used in Hermetic System). See Movable Signs.

Cazimi: A planet within 17 minutes of the Sun's center. (Not used in Hermetic System. See Report #96).

Celestial Equator: The great circle in which the plane of the earth's equator intersects the celestial sphere.

Character: Sum total of all past and present mental states; i.e., states of consciousness organized as astral energy centers in the four-dimensional form, mental factors both of the objective and unconscious mind determining the person's conduct.

Character Vibrations: Frequencies radiated by astral counterparts of physical objects including thoughts of other people. Third class of imponderable forces to which reaction takes place. See Imponderable Forces.

Chart: See Birth-chart, Natural Chart. Synonyms: Map, wheel, figure.

Chief Ruler: Strongest of several indicators.

Civil Time: Time commencing at midnight instead of noon.

Clairvoyant: One employing the psychic sense of sight.

Coeli: See Medium Coeli and Imum Coeli.

Collection of Light: Used in divinitory horary astrology; significators not within orb of aspecting each other or separating from an aspect but both forming a strong aspect to the same planet.

Colors, Astrological Signatures of: Clear colors are ruled by the planets, shades by the signs. RED: clear red, Mars; lighter red, Aries; darker red, Scorpio. ORANGE, Sun, YELLOW: clear yellow, Venus; lighter yellow, Libra; darker yellow, Taurus. GREEN, Moon. BLUE: clear blue, Saturn; lighter blue, Aquarius; darker blue, Capricorn. VIOLET: clear violet, Mercury; lighter violet, Gemini; darker violet, Virgo. PURPLE: clear purple, Jupiter; lighter purple, Sagittarius; darker purple, Pisces. Dazzling white, Uranus. Irridescence, Neptune. Infra-red, ultra-violet Pluto.

Colure: Either of two circles of the celestial sphere intersecting at the poles, one passing through the equinoctial points, the other at right angles to it.

Combust: A planet within 8 degrees and 30 minutes of the Sun. (Not used in Hermetic System. See Report #96).

Common Cross: Synonym for Mutable Cross.

Common Thought-cells: Elements of fourth-dimensional substance, mapped by zodiacal signs.

Common Signs: Synonym for Mutable Signs.

Companionship Houses, Society of: Fourth, Fifth, Sixth, Seventh.

Compartments of Astral Body: Sections of astral body mapped by twelve birth-chart houses, having an affinity for the thoughts, feelings and impulses relating to certain phases of life.

Compound: Conditioned association (mapped by an aspect) of two subconscious mental urges (mapped by the two planets in aspect).

Conciliating Planet: One harmoniously aspecting each terminal of a discordant aspect.

Conditioning: A process through which all life forms learn; education resulting from pleasureable or painful associations.

Conditioning Energy: Specific trend of desires within a thought compound.

Conjunct, verb: To form a conjunction aspect.

Conjunction Aspect: Two planets occupying the same degree of zodiacal longitude.

Conjunction Chart: See Major Conjunction Chart.

Constant: A permanent value.

Constant Logarithm: One used throughout an arithmetical process; such as one used to correct the ten planets for specific birth data.

Constellation: Irregular group of stars in outer space.

Conversion: Rearranging thought-elements within the stellar-cells and rearranging stellar-cells within their stellar-structures, without adding thought-elements of a different family.

Convertible Aspects: Parallel, Conjunction, Inconjunct.

Convertible Planets: Sun, Moon, Mercury, Uranus, Neptune, Pluto.

Co-ruler: Astrological factor sharing rulership.

Co-significator: Secondary ruler. In divinatory horary astrology; term used for Moon.

Counterpart: Replica of an object manifesting on another plane.

Cross: See Grand-square.

Culmination: Completion of a perfect aspect.

Currents of Astral Force: Energy bringing changes to man's physical body, stimulating the trend of his thinking and impressing its power upon everything by which he is surrounded.

Cusp: Dividing line between two birth-chart houses.

Cusp Ruler: Planet(s) ruling the sign appearing on a house cusp.

Cycle Charts: In mundane astrology; erected at a given locality for moment a planet crosses the celestial equator from south to north declination. One exception; Moon Cycle is erected for moment Moon is conjunction Sun (New Moon).

Cyclic Charts: Synonym for Cycle Charts.

Daily Motion of Planets: Travel in 24 hours. Daily motion varies; approximate averages follow: Sun, 59 minutes 8 seconds; Moon, 13 degrees 10 minutes; Mercury, 1 degree 23 minutes; Venus, 1 degree 12 minutes; Mars, 33 minutes 28 seconds; Jupiter, 4 minutes 59 seconds; Saturn, 2 minutes; Uranus, 42 seconds; Neptune, 22 seconds; Pluto, 15 seconds. For accurate estimate of planetary daily motion on any given day, consult ephemeris.

Daylight Time: Clocks advanced one hour.

Day Signs: Aries, Gemini, Libra, Aquarius, Sagittarius.

Decanate: Ten degrees, one-third of a zodiacal sign.

Deceleration: Decrease in planet's apparent motion relative to the earth. Synonym: Decrement.

Declination: Measurement used to locate a planet's position north or south of celestial equator.

Decreasing: Falling; referring to figures of longitude, latitude, time, etc.

Decreasing in Light: Moon in its third and fourth quarter, or its position from Full Moon until New Moon.

Decrement: See Deceleration.

Degree: A 360th part of the circumference of a circle or of a round angle. Symbol ° (as, an angle of 90°).

Degree of Combustion: Planet within 17 degrees of the Sun. (Not used in Hermetic System. See Report #96).

Degree of Emanation: Sign precedence within a zodiacal triplicity.

Degree of Exaltation: A particular degree in the sign of exaltation. See Exaltation, Sign of.

Degree of Fall: A particular degree in the sign of fall. See Fall, Sign of.

Delineation: Interpretation.

Department of Life: Birth-chart house mapping one of the twelve phases of life.

Descendant: Point, mapped by seventh house cusp, where planets sink from view on western horizon.

Descending Planet: One stationed in any house on western half of chart.

Desire Body: See Astral Body.

Desire Energy: Stimulation behind all organic activity.

Detriment, Sign of: Moderately discordant expression of the true nature of a definite subconscious urge as mapped and determined by a particular planet when located in a sign indicating unfavorable thought-cell environment for its expression.

Dignity: See Accidental Dignity, Essential Dignity.

Dimension: Level of expression: birth-chart, 2-dimensional; physical realm, 3-dimensional; astral realm, 4-dimensional; spiritual realm, 5-dimensional; celestial realm, 6-dimensional.

Directions: Term now obsolete, because ambiguously used for so many systems of progression. (Not used in Hermetic System).

Direct Planet: One apparently moving forward through zodiac.

Discordant: Predisposition toward detrimental expression.

Discordant Aspects: Semi-square, square, sesqui-square, opposition.

Discordant House: One containing more discordynes than harmodynes in a chart.

Discordant Planets: Saturn, Mars.

Discordant Sign: One containing more discordynes than harmodynes in a chart.

Discordynes: Units of astrological discord.

Diurnal: Daily.

Diurnal Revolution: Chart erected for moment Ascendant on the given day reaches the zodiacal sign, degree and minute occupied by birth-chart Ascendant.

Dominant Factor: In chart erection; precise local mean time of birth.

Dominant Planet: One having most astrodynes in a chart.

Double Rulership Signs: Scorpio, ruled by Pluto and Mars; Aquarius, by Uranus and Saturn; Pisces, by Neptune and Jupiter.

Dragon's Head and Tail: See Moon's Nodes. (Not used in Hermetic System).

Dynamic Stellar Structure: Synonymous with Dynamic Thought-Structure.

Dynamic Thought-cells: Energetic thought-cells mapped by planets.

Dynamic Thought-cell Structure: Synonymous with Dynamic Thought-Structure.

Dynamic Thought-Structure: Group of thought-cells straining to release specific desires of the planetary type mapping the structure.

Earth's Equator: An imaginary great circle on the earth's surface, everywhere equally distant from the two poles, dividing the earth's surface into the Northern and Southern Hemispheres.

Earth Signs, Triplicity of: Taurus, Virgo, Capricorn.

Eclipse: Interposition of a dark celestial body between a luminous one and the eye, or the passing of a luminous body into the shadow of another body. See Solar Eclipse and Lunar Eclipse.

Ecliptic: An elliptical path in the celestial sphere which is the apparent path of the sun, or of the earth as seen from the sun, where eclipses of sun ànd moon take place.

Eighth House: Segment of chart mapping the particular section of man's astral body, or world's astral form, where inner-plane activities determine events concerning death, inheritance, taxes and other people's money, among other things.

Electric Planets: Sun, Mars, Jupiter.

Electromagnetic: Adjective describing electric and magnetic energies having the velocity of light, making possible transition of messages from higher-velocity astral plane to lower-velocity physical plane, or vice versa. Used with such words as body, form, belt, boundary or region.

Elements: Fire, Earth, Air, Water; referring to zodiacal signs.

Elevated Planet: One stationed near apex of birth-chart, or at least in one of the six houses above horizon.

Eleventh House: Segment of chart mapping the particular section of man's astral body, or world's astral form, where inner-plane activities determine events concerning friends, hopes, and acquaintances, among other things.

Energy-release: Expression of potential mapped by a progressed aspect within one degree of perfect.

Environment: Aggregate of all external conditions and influences affecting life and development of an organism on any plane—physical, astral, spiritual, celestial.

Ephemeris: Tabular statement of assigned places in zodiac of planets for regular intervals. Usually covers one year. Plural: ephemerides.

Epoch: Event marking the beginning of a new development.

Equator: See Earth's Equator and Celestial Equator.

Equinoctial Colure: Intercepted point of the ecliptic and celestial equator.

Equinox: Times when nights are everywhere equal in duration to days. See Autumnal Equinox and Vernal Equinox.

Equivalent Greenwich Mean Time: Time at Greenwich in reference to local mean time at some locality on earth.

Equivalent Greenwich Mean Time Interval: The number of hours, minutes and seconds from a Greenwich time reference: if after the time reference, called a Plus Interval; if before, called a Minus Interval.

Equivalent Local Mean Time: Local mean time at some locality on earth in reference to time at Greenwich.

Essential Dignity: Harmonious or discordant expression of the true nature of a definite subconscious urge as mapped and determined by a particular planet when located in a sign indicating favorable or unfavorable thought-cell environment for its expression.

Event: That which comes, arrives, or happens. Occurrence.

Exaltation, Sign of: Highly harmonious expression of the true nature of a definite subconscious urge as mapped and determined by a particular planet when located in a sign indicating favorable thought-cell environment for its expression.

Expansion Compound: Conditioned association (mapped by the inconjunct aspect) of two subconscious urges (mapped by the two planets involved in the aspect).

Extrasensory Perception: Ability to gain information regarding past, present or future with means other than the physical senses.

Fall Chart: Erected for moment Sun enters Libra, or for autumnal equinox.

Fall, Sign of: Highly discordant expression of the true nature of a definite subconscious urge as mapped and determined by a particular planet when located in a sign indicating unfavorable thought-cell environment for its expression.

Fast Ascendant: Travel within a 24-hour period exceeding travel of Midheaven for same period.

Fast Planet: Travel in excess of normal motion. See Daily Motion of Planet.

Favorable: See Harmonious.

Feeble Houses: Third, Second.

Feminine Planets: Moon, Venus.

Feminine Signs: Taurus, Cancer, Virgo, Scorpio, Capricorn, Pisces.

Fifth House: Segment of chart mapping the particular section of man's astral body, or world's astral form, where inner-plane activities determine events concerning speculation, children, love affairs and entertainment, among other things.

Fire Signs, Triplicity of: Aries, Leo, Sagittarius.

First House: Segment of chart mapping the particular section of man's astral body, or world's astral form, where inner-plane activities determine events concerning personality, health and physical body, among other things.

First Key Problem: In chart erection; finding the sidereal time of birth.

First Quarter of Moon: Period covering time Moon conjuncts Sun until Moon squares Sun.

Fixed Cross: Grand-square in fixed signs.

Fixed Signs, Quality of: Taurus, Leo, Scorpio, Aquarius.

Fixed Star: One outside of our solar system. (Not used in Hermetic System. See Report #81.)

Fortunate: See Harmonious.

Fourth House: Segment of chart mapping the particular section of man's astral body, or world's astral form, where inner-plane activities determine events concerning father, real estate and home, among other things.

Fourth Quarter of Moon: Period covering time Moon squares Sun (second square) until Moon conjuncts Sun.

Friction Compound: Conditioned association (mapped by the semi-square aspect) of two subconscious urges (mapped by the two planets involved in the aspect).

Fruitful Signs: Cancer, Scorpio, Pisces. Moderately fruitful: Taurus,

Gemini, Virgo, Libra, Sagittarius, Aquarius.

Full Moon: Moon opposition Sun: begins third quarter of Moon.

Gemini: Third sign of zodiac. "I think" motivation. Positive thought-cell activity, versatility; negative, changeableness.

Geocentric: Relating to or measured from the earth's center.

Grand-Cross: See Grand-square.

Grand-Square: Two planets in opposition both squaring two other planets in opposition.

Grand-trine: Two planets in trine aspect both trine a third planet.

Greater Benefic: Jupiter.

Greater Malefic: Saturn.

Greenwich: Place in England intersected by 0° meridian of longitude.

Greenwich Civil Time: Time commencing at midnight, or 0 hours.

Greenwich Mean Time: Local mean time at Greenwich.

Groundwire: Term used for Ascendant mapping the point where native connects with his immediate environment and exchanges energy through the electromagnetic body and its forces with that environment.

Growth Compound: Conditioned association (mapped by the semi-sextile aspect) of two subconscious mental urges (mapped by the two planets involved in the aspect).

Harmodynes: Units of astrological harmony.

Harmonious: Predisposition to express benevolently.

Harmonious Aspects: Semi-sextile, sextile, trine.

Harmonious House: One containing more harmodynes than discordynes.

Harmonious Planets: Jupiter, Venus.

Harmonious Sign: One having more harmodynes than discordynes.

Harmony, Sign of: Slightly harmonious expression of the true nature of a definite subconscious urge as mapped and determined by a particular planet when located in a sign indicating favorable thought-cell environment for its expression.

Hermetic Science: Science of eternal mind relating to physical, astral and spiritual matters. Hermetic was derived from Hermes - ancient name for Mercury - who had wings on his feet (understanding) and was called the "messenger of the gods." Hermetic also signifies "forever", just as hermetically sealed means "forever sealed or closed"; showing relationship to eternity.

Hermetic System of Astrology: Source material presented in the 21 Brotherhood of Light Courses, issued by the Church of Light.

Higher Octave Planets: Uranus, Neptune, Pluto.

Home Sign: One which the particular planet rules. Moderately harmonious expression of the true nature of a definite subconscious urge as mapped and determined by a particular planet when located in a sign indicating favorable thought-cell environment for its expression.

Horary Astrology: One of the seven branches of astrology used to answer questions. When a question becomes clear and is expressed verbally or in writing, a chart is erected for place, date and precise moment where the question is given birth. Its function is to explore the possibilities of the question and what will transpire relative to it in the future.

Horizon: Horizontal line bisecting circle.

Horoscope: Chart erected for a definite date, place and time.

Hour Ruler: Planet ruling a definite period of time within 24 hours.

House: Segment of chart mapping the particular section of man's astral body, or particular section of the world's astral form, the inner-plane activities of which determine events attracted in each department of life.

House Cusp: Dividing line between two houses.

House Power: Power of planet due to house position alone, designated in astrodynes.

House Rulership: Planet(s) ruling house. (See Report #95).

Imponderable Forces: Three general categories of astral frequencies: radiations from mental processes, astral counterparts of physical objects, and planets. See Astrological Vibrations, Character Vibrations, Thought Vibrations.

Imum Coeli: Latin equivalent of Nadir.

Inconjunct Aspect: Two planets occupying positions 150° apart in zodiacal longitude.

Increasing: Rising; referring to figures of latitude, longitude, time, etc.

Increasing in Light: Moon in its first and second quarter, or its position from New Moon to Full Moon.

Increment: See Acceleration.

Influence: Term loosely used in astrology. A person's condition in life is not caused by astrological vibrations. Planets radiate habitual frequencies corresponding in keytone to centers of energy (basic urges) within the subconscious mind. Mental and emotional REACTION to astrological and other energies produce events. Therefore, the cause of events is not the planets but reactions to stimuli determined by conditioning energy — painful or pleasureable — associated with prior reaction.

Ingress Map: Term commonly used for Sun Cycle Chart.

Inharmonious: See Discordant.

Inharmony, Sign of: Slightly discordant expression of the true nature of a definite subconscious urge as mapped and determined by a particular planet when located in a sign indicating unfavorable thought-cell environment for its expression.

Inner-plane: See Astral Plane.

Intellectual Planets: Mercury, Uranus.

Intensity Compound: Conditioned association (mapped by the parallel aspect) of two subconscious urges (mapped by the two planets involved in the aspect).

Intercepted Sign: One appearing between house cusps, not on cusp.

Interception. Referring to intercepted sign.

Interval: Period before or after a given time reference in hours, minutes and seconds, or in days, months and years.

Involved (ing): Word referring to planets in aspect, instead of repeating the phrase "making an aspect to or receiving an aspect from."

Jupiter: Planetary symbol equivalent to subconscious religious urge. Thought-cell activity: general, abundance; positive, benevolence; negative, conceit.

Kinetic: Of, pertaining to, or due to changes of motion or form produced by forces.

Latitude: Angular distance measured in a meridian; distance measured in degrees (degrees of latitude) north or south, from the equator.

Leo: Fifth sign of zodiac. "I will" motivation. Positive thought-cell activity, kindness; negative, domination.

Lesser Benefic: Venus.

Lesser Malefic: Mars.

Libra: Seventh sign of zodiac: "I balance" motivation. Positive thought-cell activity, affability; negative, approbation.

Life Houses, Trinity of: First, Fifth, Ninth.

Limiting Date: A constant enabling calculation of time within the particular calendar year a major progressed aspect becomes perfect.

Local Mean Time: Local time measured by the apparent westward motion of the mean sun, but actually due to the uniform eastward turning of the earth on its axis.

Local Mean Time Interval: Period of time between the local mean time and some time reference; if after the reference point, called a Plus Interval; if before, a Minus Interval.

Logarithms: Mathematical expedient, eliminating multiplication and division. Proportional diurnal logarithms used in chart calculations.

Longitude: Arc or portion of the equator intersected between meridian of a given place and prime meridian, as East or West of Greenwich. Longitude of a place expressed either in degrees (longitude in arc) or in time (longitude in time).

Lower Octave Planets: Sun, Moon, Mercury, Venus, Mars, Jupiter, Saturn.

Luminary: Sun, Moon. To avoid repeating "the luminaries and the planets", the luminaries are called planets.

Luck Compound: Conditioned association (mapped by the trine aspect) of two subconscious mental urges (mapped by the two planets involved in the aspect).

Lunar: Referring to Moon.

Lunar Constant: A permanent value enabling calculation of time within the particular calendar month a minor progressed aspect becomes perfect.

Lunar Eclipse: Moon entering the earth's shadow.

Lunar Revolution: Two types; chart erected for moment transiting Moon (1) conjuncts its birth-chart place, or (2) conjuncts natal Sun. (See Report #83).

Lunation Map: Term commonly used for Moon Cycle Chart.

m: Abbreviation for minor progressed position.

Magnetic Planets: Moon, Venus, Saturn, Neptune, Pluto.

Major Conjunction Chart: Map erected for some location at the exact time two major planets complete a conjunction.

Major Progressions: Energy-release subsequent to birth calculated from a day in the ephemeris equaling a calendar year in life.

Major Progression Date: A day in the ephemeris showing planetary positions by major progression for some day during the calendar year.

Major Progression Time: Period of storage for mapped energy released by major progressions.

Malefic: See Discordant.

Map: Synonym for chart.

Mars: Planetary symbol equivalent to subconscious aggressive urge. Thought-cell activity: general, strife; positive, initiative; negative, harshness.

Masculine Planets: Sun, Mars, Jupiter, Saturn.

Masculine Signs: Aries, Gemini, Leo, Libra, Sagittarius, Aquarius.

Mean Time: Time measured by apparent westward motion of the mean sun.

Medium Coeli: Latin equivalent to Midheaven.

Mental Antidote: Thought remedy to counteract the effects of mental poisoning as mapped by a discordant planet. (See Report #97).

Mental Compound: An association—either harmonious or discordant—of two groups of thought-elements mapped by an aspect.

Mental Planets: Mercury, Uranus, Moon.

Mercury: Planetary symbol equivalent to subconscious intellectual urge. Thought-cell activity: general, thought; positive, expression; negative, restlessness.

Meridian: A great circle of the celestial sphere passing through its poles and the zenith of a given place, a series of such lines being numbered according to degrees of longitude.

Midheaven: Point directly overhead; zenith; 10th house cusp.

Midheaven Constant: Permanent value of distance between natal Sun and Midheaven in a birth-chart. Used to calculate major, minor or transit progressed Midheaven.

Midnight Ephemeris: More properly, Zero Hour Ephemeris. Calculated for 0 hours. See Ephemeris.

Minor Ephemeris Date: A day in the

ephemeris showing planetary positions of minor progressions for some day during the calendar year.

Minor House Cusps: All other house cusps aside from Midheaven (10th) and Ascendant (1st).

Minor Progressions: Energy-release subsequent to birth calculated from a month in the ephemeris equaling a calendar year in life.

Minor Progression Date: A day in the ephemeris showing planetary positions by minor progression for birthday during the calendar year.

Minor Progression Time: Period of storage for mapped energy released by minor progressions.

Minus Interval: A period of time before some time reference point.

Moderately Powerful Houses: Eleventh, Eighth, Ninth, Twelfth.

Moisture Chart: Used in weather predicting; erected for moment Moon enters each of its four quarters.

Moon: Planetary symbol equivalent to subconscious domestic urge. Thought-cell activity: general, fluctutation; positive, adaptability; negative, inconstancy.

Moon Cycle Chart: Erected for moment Moon conjuncts Sun. Synonyms: Lunation, New Moon Chart.

Moon Phases: See First, Second, Third, Fourth Quarter of Moon.

Moon-Sign: Zodiacal location of Moon.

Moon's Nodes: Two points of intersection of Moon's orbit and earth's orbit. (Not used in Hermetic System. See Report #81.)

Movable Cross: Grand-square in movable signs.

Movable Signs, Quality of: Aries, Cancer, Libra, Capricorn.

Mundane Astrology: One of the seven branches of astrology based on Cycle Charts and charts of nations, cities, communities, and other groups of people under a common governing authority.

Mundane House: One segment of the circle about the earth from east to west, mapping one of the 12 departments of life, or 12 compartments of the astral body or world's astral form.

Mutable Cross: Grand-square in mutable signs.

Mutable Signs, Quality of: Gemini, Virgo, Sagittarius, Pisces.

Mutual Reception: Two planets, not necessarily in aspect to each other, occupying the exaltation or home sign of each other.

Nadir: Lowest point in birth-chart, mapped by 4th house cusp.

Natal: Adjective referring to birth-chart positions.

Natal Astrology: One of the seven branches of astrology dealing with the character of people, groups of people under a single governing authority, enterprises, life-forms other than man, and other things which have a definite moment of birth; and with their reactions to inner-plane weather.

Native: Person for whom chart is erected.

Nativity: See Birth-chart.

Natural Chart: One with Aries on 1st house cusp, Taurus on 2nd, etc., around the zodiac.

Negative Planets: Moon, Venus, Saturn, Neptune, Pluto.

Negative Signs: Taurus, Cancer, Virgo, Scorpio, Capricorn, Pisces.

Neptune: Planetary symbol equivalent to subconscious utopian urge. Thought-cell activity: general, illusion; positive, idealism; negative, vagueness.

Neutral Aspects: See Convertible Aspects.

Neutral Planets: Mercury, Uranus, Neptune, Pluto.

New Moon: Moon conjunction Sun; beginning first quarter Moon.

New Moon Chart: Map erected for precise moment Moon conjuncts Sun.

Night Signs: Taurus, Virgo, Scorpio, Capricorn, Pisces.

Ninth House: Segment of chart mapping the particular section of man's astral body, or world's astral form, where inner-plane activities determine events concerning publicly expressed opinions, books, religion and long journeys, among other things.

Nodes: Intersection of the orbital paths of two heavenly bodies.

Normal: Predispositions mapped by birth-chart.

Obstacle Compound: Conditioned association (mapped by the square aspect) of two subconscious mental urges (mapped by the two planets involved in the aspect).

Opportunity Compound: Conditioned association (mapped by the sextile aspect) of two subconscious mental urges (mapped by the two planets involved in the aspect).

Opposition Aspect: Two planets occupying positions 180° apart in zodiacal longitude.

Operative: Within orb of aspect: natal, major, minor or transit progressions.

Orb: Number of degrees allowed before or after a perfect aspect. (See Orb Table and Report #31.)

Outer-Plane: Physical environment.

Overlapping Aspects: Semi-square (45°) and sesqui-square (135°); so called because their angles do not extend in 30° segments.

p: Abbreviation for major progressed position.

Parallel Aspect: Two planets occupying same degree, north or south, of declination.

Partile: Obsolete word for perfect.

Part of Fortune: Point on ecliptic as far removed from Ascendant by longitude as Moon is removed from Sun by longitude. (Not used in Hermetic System. See Report #81.)

Past Aspect: Planet(s) beyond perfect aspect.

Peak Power: Periods during an active major progressed aspect mapped by progressed Moon, a minor progression, or both, or several minor progressions aspecting one of major planets involved. Period of unusual thought-cell activity.

Perfect: Exact.

Permanent Aerial: Line across astral body mapped by a birth-chart aspect.

Permanent Stellar Aerial: Synonymous with Permanent Aerial.

Personal Houses, Trinity of: Twelfth, First, Second, Third.

Pisces: Twelfth sign of zodiac. "I believe" motivation. Positive thought-cell activity, sympathy; negative, worry.

Planetary: Adjective referring to planet.

Planetary Family: All those things on any plane having the astrological signature of a particular planet; ten families, each corresponding to a planet.

Planetary Hour: One measuring exactly one-twelfth of the time from sunrise to sunset if a day hour; and one-twelfth the time from sunset to sunrise if a night hour.

Planetary Nodes: See Nodes.

Planets: Symbolic equivalents to dynamic subconscious urges.

Plus Interval: Period of time following some time reference.

Pluto: Planetary symbol equivalent to subconscious universal welfare urge. Thought-cell activity: general, coercion; positive, spirituality; negative, inversion.

Positive Planets: Sun, Mars, Jupiter.

Positive Signs: Aries, Gemini, Leo, Libra, Sagittarius, Aquarius.

Powerful Houses: First, Tenth, Seventh, Fourth.

Precautionary Actions: Measures taken beforehand to ward off discord or to secure harmony or success.

Predisposition. Tendency.

Predominant: Having ascendancy over others, due to strength, influence, etc.

Prenatal Epoch: Moment of conception (not the act). Theory of the relation between time of conception and time of birth. (See Report #94.)

Prime Meridian: A meridian intersecting the equator from which longitude is counted, both east and west. That of Greenwich is almost universally used.

Progressed (sion): Referring to aspects and positions of major, minor and transit energy-releases.

Progressed Aspect: Temporary stellar aerial between dynamic thought-structures acquiring sufficient psychokinetic energy to attract events of a corresponding nature.

Progressed Constant: Permanent value of a progression.

Prominent Planet: Specific: having many astrodynes. General: in an anglé; aspecting Sun, Moon or Mercury; or involved in many aspects.

Prominence Compound: Conditioned association (mapped by the conjunction aspect) of two subconscious mental urges (mapped by the two planets involved in the aspect).

Psychic Houses, Trinity of: Fourth, Eighth, Twelfth.

Psychokinetic: Of, pertaining to, or due to changes of motion or form produced by non-physical forces.

Public Houses: Eighth, Ninth, Tenth, Eleventh.

Quadrant: Quarter of a circle; bounded by a quadrant and two radii.

Qualities, Zodiacal: Three in number: movable, fixed, mutable signs.

Querent: Used in horary astrology: person asking question.

Quesited: Used in horary astrology; thing asked about.

r: Abbreviation for birth-chart positions.

Radical Figure: Used in horary astrology; map erected for exact time of commencement of an enterprise or event.

Radix: See Birth-chart.

Rallying Forces: Energy derived from stimulated thought-cell activity, mapped by total progressed aspects.

Reenforcement Effect: Minor progressions adding accessory energy to a major progression. (See Report #23.)

Retrograde Planet: One apparently moving backward through the zodiac, due to the earth's relative motion.

Right Ascension: Distance east or west of 0° Aries expressed in degrees, minutes and seconds. (Not used in Hermetic System).

Rising Degree: Degree on Ascendant.

Rising Planet: One stationed in any house on eastern side of chart.

Rising-Sign: Sign on Ascendant.

Ruler: General, an astrological signature vibrating in sympathetic response to a manifestation on any plane—physical, astral or spiritual.

Ruler of Houses: See House Rulership.

Ruler of Signs: Single planetary rulership: Aries by Mars; Taurus, Venus; Gemini, Mercury; Cancer, Moon; Leo, Sun; Virgo, Mercury; Libra, Venus; Sagittarius, Jupiter; Capricorn, Saturn. Double planetary rulership: Scorpio, Pluto and Mars; Aquarius, Uranus and Saturn; Pisces, Neptune and Jupiter.

Ruling Planet of Chart: Dominant planet in a chart, one having most astrodynes.

Sagittarius: Ninth sign of zodiac. "I see" motivation. Positive thought-cell activity, loyalty; negative, sportiveness.

Saturn: Planetary symbol eqivalent to subconscious safety urge. Thought-cell activity: general, poverty; positive, system; negative, selfishness.

Scorpio: Eighth sign of zodiac. "I desire" motivation. Positive thought-cell activity, resourcefulness; negative, troublesomeness.

Second House: Segment of chart mapping the particular section of man's astral body, or world's astral form, where inner-plane activities determine events concerning money and personal property, among other things.

Second Key Problem: In chart erection; finding the Equivalent Greenwich Mean Time Interval of birth.

Second Quarter of Moon: Period covering time Moon squares Sun until Moon opposes Sun.

Semi-sextile Aspect: Two planets occupying positions 30° apart in zodiacal longitude.

Semi-square Aspect: Two planets occupying positions 45° apart in zodiacal longitude.

Separating Aspect: See Past Aspect.

Separation Compound: Conditioned association (mapped by the opposition aspect) of two subconscious mental urges (mapped by the two planets involved in the aspect).

Sesqui-square Aspect: Two planets occupying positions 135° apart in zodiacal longitude.

Seventh House: Segment of chart mapping the particular section of man's astral body, or world's astral form, where inner-plane activities determine events concerning marriage, partnership, the public and open enemies, among other things.

Sextile Aspect: Two planets occupying positions 60° apart in zodiacal longitude.

Sidereal Time: Distance along zodiacal belt from 0° Aries expressed in hours, minutes and seconds.

Sidereal Time of Birth: Distance of a specific point in time along zodiacal belt from 0° Aries. Used to calculate signs on house cusps.

Significator: Planet ruling house denoting any person or thing.

Sign: 30° segment of zodiacal belt, mapping motivation behind action.

Sixth House: Segment of chart mapping the particular section of man's astral body, or world's astral form, where inner-plane activities determine events concerning work, illness, food, and employees, among other things.

Slow Ascendant: Travel within 24 hours less than travel of Midheaven for same period.

Slow Planet: One traveling less than normal motion. See Daily Motion of Planet.

Social Planets: Venus, Mars, Neptune, Pluto.

Societies, House: Three. See Personal, Companionship, Public Houses.

Solar: Referring to Sun.

Solar Constant: Permanent birth-chart value of distance between Moon and Sun. Used to calculate minor progressions.

Solar Eclipse: Passing of moon between sun and earth.

Solar Revolution: Chart erected for moment transiting Sun returns to sign, degree and minute of birth-chart Sun. (See Report #84).

Solstice: Point in ecliptic at which sun is farthest from equator, north or south, namely first point of Cancer (summer solstice) and first point of Capricorn (winter solstice).

Spiritual Astrology: One of the seven branches of astrology embracing the relationship between astrological factors and Divine Law and Order, including moral conceptions and customs, as observed by wise men throughout the ages.

Spring Chart: Erected for moment Sun enteres Aries.

Square Aspect: Two planets occupying positions 90° apart in zodiacal longitude.

Standard Time: Civil time established by law or by general usage over a region, corresponding to mean local time of a meridian.

Standard Time Meridian: Local mean time at meridian determining the clock time in its region.

Standard Time Zone: Area adopting local mean time of a particular meridian.

Standardized Symbols of Aspects:

☌ Conjunction	△ Trine
⚺ Semi-sextile	⚼ Sesqui-Square
∠ Semi-square	⚻ Inconjunct
⚹ Sextile	☍ Opposition
□ Square	P Parallel

Standardized Symbols of Planets:

☉ Sun	♃ Jupiter
☽ Moon	♄ Saturn
☿ Mercury	♅ Uranus
♀ Venus	♆ Neptune
♂ Mars	♇ Pluto

Standardized Symbols of Zodiacal Signs:

♈ Aries	♉ Taurus	♊ Gemini	♋ Cancer
♌ Leo	♍ Virgo	♎ Libra	♏ Scorpio
♐ Sagittarius	♑ Capricorn	♒ Aquarius	♓ Pisces

Station: House position of planet.

Stationary Planet: A direct planet turning retrograde, or a retrograde planet turning direct in motion, for several days apparently having no zodiacal motion.

Stellar Aerials: Lines across astral body picking up, radio fashion, planetary energy-releases. Mapped by aspects.

Stellar Cells: See Astral Cells.

Stellar Diagnosis: One of the seven branches of astrology, a specialized section of natal astrology, relating to health.

Stellar Dynamics: Method of gauging power (astrodynes), harmony (harmodynes) and discord (discordynes) of astrological factors in one chart (frame of reference).

Stellar Healing: Treatment of astral body to heal physical.

Stellar Structure: Formation in stellar anatomy; energy-center mapped by a planet.

Stimulate (ed): Step up in activity.

Strongest Planet: See Dominant Planet.

Subconscious Mind: See Unconscious Mind.

Sublimation: Finding an acceptable channel for release of the essential nature of a desire.

Sub-Major Progression: Term used for major progressed Moon aspects.

Succeedent Houses: Second, Fifth, Eighth, Eleventh. Houses of moderate volume.

Summer Chart: Map erected for moment Sun enters Cancer.

Summer Solstice: See Solstice.

Sun: Planetary symbol equivalent to subconscious power urge. Thought-cell activity: general, vigor; positive, rulership; negative, dictativeness.

Sun Cycle Chart: Map erected for moment Sun enters Aries.

Sun-Sign: Zodiacal location of Sun.

t: Abbreviation for transit progressed position.

Table of Houses: Data for determining relation of heavens to a particular place on earth at a given time. Used in calculation of house cusps.

Taurus: Second sign of zodiac. "I have" motivation. Positive thought-cell activity, stability; negative, obstinacy.

Temperature Chart: Maps erected for four seasons; moment Sun enters Aries, Cancer, Libra, Capricorn.

Temporary Stellar Aerial: Line across astral body picking up, radio fashion, energy-releases for a limited time. Mapped by a progressed aspect.

Tenth House: Segment of chart mapping the particular section of man's astral body, or world's astral form, where inner-plane activities determine events concerning the job, business, honor and reputation, among other things.

Terminal: Point at each end of an aerial, mapped by a birth-chart or progressed planet.

Third House: Segment of chart mapping the particular section of man's astral body, or world's astral form, where inner-plane activities determine events concerning thoughts, studies, short journeys and relatives, among other things.

Third Quarter of Moon: Period covering time Moon opposes Sun until Moon squares Sun.

Thought-cells: Elements of astral substance, types determined by astrological equivalents. See Astral Cells.

Thought-cell Structure: Colony of cells, having the same desire, mapped by a planet.

Thought Compound: Combination of two thought structures, mapped by an aspect.

Thought Family: A colony of cells, straining to express the same desire, mapped by a planet.

Thought Structure: See Thought-cell Structure.

Thought Vibrations: Frequencies radiated by an individual's own thoughts. Most important class of imponderable forces to which reaction takes place. See Imponderable Forces.

Time Interval: Period of time expressed in hours, minutes and seconds, or in years, months and days.

Transit: Referring to current planetary position in the sky.

Transit Date: A day in ephemeris showing planetary positions by transit progression for some day during calendar year.

Translation of Light: Used in divinatory horary astrology: significators separating from an aspect, and some other planet forming an aspect to one significator and moving to a complete aspect with the other significator.

Trigger Effect: Referring to release of energies by a transit progression. (See Report #91.)

Trine Aspect: Two planets occupying positions 120° apart in zodiacal longitude.

Trinity of Houses: Four in number: Life, Wealth, Association, Psychism.

Triplicity, Zodiacal: Four in number, each relating to an element: fire, earth, air, water.

True Sidereal Time: See Sidereal Time of Birth.

T-square: Two planets in opposition squaring a third planet.

Twelfth House: Segment of chart mapping the particular section of man's astral body, or world's astral form, where inner-plane activities determine events concerning secret enemies, disappointments, and astral entities, among other things.

Unconscious Mind: A non-physical organization of thought-elements into thought-cells and thought-structures, storing the sum total of all states of consciousness.

Unfortunate: See Discordant.

Ununiform Motion of Planets: Planet's apparent speed of travel varying at different times during a 24-hour period, accelerating or decelerating as planet increases distance from midnight.

Upper Octave Planet: Uranus, Neptune, Pluto. (See Report #90.)

Uranus: Planetary symbol equivalent to subconscious individualistic urge. Thought-cell activity: general, unconventionality; positive, originality; negative, eccentricity.

Urge: An organization of thought-cells into a thought-structure in the subconscious mind possessing a powerful desire, mapped by a planet.

Venus: Planetary symbol equivalent to subconscious social urge. Thought-cell activity: general, gratuities; positive, affection; negative, pliancy.

Vernal Equinox: Point where sun apparently following the ecliptic crosses the celestial equator in Spring.

Vernal Ingress: Synonymous with Sun Cycle Chart.

Vibrations: Frequencies with a basic key-tone, or beat note.

Virgo: Sixth sign of zodiac. "I analyze" motivation. Positive thought-cell activity, analysis; negative, criticism.

Vital Planets: Sun, Moon.

Vocational Astrology: Branch of natal astrology analyzing birth-chart constants for vocations.

Void of Course: Used in divinatory horary astrology; a planet making no aspect with another planet before passing from sign it is in.

Waxing and Waning: Referring to Moon decreasing and increasing in light.

Water Signs, Triplicity of: Cancer, Scorpio, Pisces.

Weakest Planet: One having least astrodynes in a chart.

Wealth Houses, Trinity of: Second, Sixth, Tenth.

Weather Predicting: One of the seven branches of astrology related to the nature and changes in weather employing temperature charts, wind charts and moisture charts.

Wind Chart: Used in weather predicting; erected for moment Mercury, by either direct or retrograde motion, enters a new zodiacal sign.

Winter Chart: Map erected for moment Sun enters Capricorn.

Winter Solstice: See Solstice.

Worst Aspect: One having most discordynes in a chart.

Worst House: One containing most discordynes in a chart.

Worst Planet: One having most discordynes in a chart.

Worst Sign: One having most discordynes in a chart.

Zenith: Point directly overhead in sky, mapped in a birth-chart by 10th house cusp, or Midheaven.

Zero Hour (Midnight) Ephemeris: Planetary positions at zero hours given for each day of the year. See Ephemeris.

Zodiac: A circular belt in the sky containing twelve equal segments.

Zodiacal: Adjective, referring to zodiac.

Zone: Part of astral body, mapped by a zodiacal sign.

INDEX